T0185258

Virtual Crowds

Steps Toward Behavioral Realism

Synthesis Lectures on Visual Computing

Computer Graphics, Animation, Computational Photography, and Imaging

Editor
Brian A. Barsky, *University of California, Berkeley*

This series presents lectures on research and development in visual computing for an audience of professional developers, researchers and advanced students. Topics of interest include computational photography, animation, visualization, special effects, game design, image techniques, computational geometry, modeling, rendering, and others of interest to the visual computing system developer or researcher.

Virtual Crowds: Steps Toward Behavioral Realism
Mubbasir Kapadia, Nuria Pelechano, Jan Allbeck, and Norm Badler
2015

Finite Element Method Simulation of 3D Deformable Solids
Eftychios Sifakis and Jernej Barbič
2015

Efficient Quadrature Rules for Illumination Integrals: From Quasi Monte Carlo to Bayesian Monte Carlo
Ricardo Marques, Christian Bouville, Luís Paulo Santos, and Kadi Bouatouch
2015

Numerical Methods for Linear Complementarity Problems in Physics-Based Animation
Sarah Niebe and Kenny Erleben
2015

Mathematical Basics of Motion and Deformation in Computer Graphics
Ken Anjyo and Hiroyuki Ochiai
2014

Mathematical Tools for Shape Analysis and Description
Silvia Biasotti, Bianca Falcidieno, Daniela Giorgi, and Michela Spagnuolo
2014

Information Theory Tools for Image Processing
Miquel Feixas, Anton Bardera, Jaume Rigau, Qing Xu, and Mateu Sbert
2014

Gazing at Games: An Introduction to Eye Tracking Control
Veronica Sundstedt
2012

Rethinking Quaternions
Ron Goldman
2010

Information Theory Tools for Computer Graphics
Mateu Sbert, Miquel Feixas, Jaume Rigau, Miguel Chover, and Ivan Viola
2009

Introductory Tiling Theory for Computer Graphics
Craig S.Kaplan
2009

Practical Global Illumination with Irradiance Caching
Jaroslav Krivanek and Pascal Gautron
2009

Wang Tiles in Computer Graphics
Ares Lagae
2009

Virtual Crowds: Methods, Simulation, and Control
Nuria Pelechano, Jan M. Allbeck, and Norman I. Badler
2008

Interactive Shape Design
Marie-Paule Cani, Takeo Igarashi, and Geoff Wyvill
2008

Real-Time Massive Model Rendering
Sung-eui Yoon, Enrico Gobbetti, David Kasik, and Dinesh Manocha
2008

High Dynamic Range Video
Karol Myszkowski, Rafal Mantiuk, and Grzegorz Krawczyk
2008

GPU-Based Techniques for Global Illumination Effects
László Szirmay-Kalos, László Szécsi, and Mateu Sbert
2008

High Dynamic Range Image Reconstruction
Asla M. Sá, Paulo Cezar Carvalho, and Luiz Velho
2008

High Fidelity Haptic Rendering
Miguel A. Otaduy and Ming C. Lin
2006

A Blossoming Development of Splines
Stephen Mann
2006

Virtual Crowds: Steps Toward Behavioral Realism
Mubbasir Kapadia, Nuria Pelechano, Jan Allbeck, and Norm Badler

ISBN: 978-3-031-01458-1 paperback
ISBN: 978-3-031-02586-0 ebook

DOI 10.1007/978-3-031-02586-0

A Publication in the Springer series
SYNTHESIS LECTURES ON VISUAL COMPUTING: COMPUTER GRAPHICS, ANIMATION, COMPUTATIONAL PHOTOGRAPHY, AND IMAGING

Lecture #20
Series Editor: Brian A. Barsky, *University of California, Berkeley*
Series ISSN
ISSN pending.

Virtual Crowds

Steps Toward Behavioral Realism

Mubbasir Kapadia
Rutgers University

Nuria Pelechano
Universitat Politècnica de Catalunya

Jan Allbeck
George Mason University

Norm Badler
University of Pennsylvania

SYNTHESIS LECTURES ON VISUAL COMPUTING: COMPUTER GRAPHICS, ANIMATION, COMPUTATIONAL PHOTOGRAPHY, AND IMAGING #20

ABSTRACT

This volume presents novel computational models for representing digital humans and their interactions with other virtual characters and meaningful environments. In this context, we describe efficient algorithms to animate, control, and author human-like agents having their own set of unique capabilities, personalities, and desires. We begin with the lowest level of footstep determination to steer agents in collision-free paths. Steering choices are controlled by navigation in complex environments, including multi-domain planning with dynamically changing situations. Virtual agents are given perceptual capabilities analogous to those of real people, including sound perception, multi-sense attention, and understanding of environment semantics which affect their behavior choices. The roles and impacts of individual attributes, such as memory and personality are explored. The animation challenges of integrating a number of simultaneous behavior and movement demands on an agent are addressed through an open source software system. Finally, the creation of stories and narratives with groups of agents subject to planning and environmental constraints culminates the presentation.

KEYWORDS

computer graphics, crowd simulation, computer animation, agent simulation, steering, navigation, semantic modeling, agent perception, sound, attention, behavior selection, narrative, digital storytelling, pathfinding, behavior authoring

To our families, friends, colleagues, and students

Contents

Preface

In 2008 Nuria Pelechano, Jan Allbeck, and Norm Badler proposed to author a Synthesis Lecture on Computer Graphics and Animation for Morgan and Claypool Publishers, entitled *Virtual Crowds: Methods, Simulation, and Control.* The field of crowd simulation was hardly new, but it had a relatively scattered range of approaches and publications which we felt demanded some organization, explication, and development. During the writing process we were fortunate to construct the volume from foundational crowd simulation work in the PhD theses of Pelechano and Allbeck. The overarching principle we followed was the "bottom-to-top" nature of virtual crowd animation methods: the gamut of approaches extending from low-level collision-avoidance control methods to high-level behaviors based on agent schedules, needs, and reactions to situations.

Seven years later, the need, even urge, to update this work became paramount. The crowd simulation community was increasing rapidly in number of researchers, practitioners, users, and their publications. We re-assembled the Pelechano, Allbeck, and Badler team from the 2008 volume, and added a new co-author, Mubbasir Kapadia. Kapadia had worked as a postdoc in Badler's University of Pennsylvania lab for almost two years, and was well equipped to help bring the work up-to-date in a new volume.

As the four of us discussed how to structure the new book, two principles were easily established at the outset. First, the new book would not be just a re-edit and update of the material in the 2008 volume. The present version is indeed a totally new work. However, the second principle extended what was already established in the first volume: the material would again cover the lowest level of steering and navigation control and build up to the highest possible levels of group activity specification. In fact, the current volume goes all the way to narrative storytelling and planning. These two principles lead into new topics, such as automatically computing navigable spaces from geometry models, and adding new sensory capabilities to guide agent reactions and behaviors.

Our approach to vertical exposition leaves aside much parallel work by others in the fields of crowd simulation, computer graphics character modeling, cognitive agents, and specific application domains, such as evacuation models. In return, however, we integrate a considerable amount of dispersed technical material that, if nothing else, will help endow future virtual crowd research with a more global world view of the crowd simulation field.

Mubbasir Kapadia, Nuria Pelechano, Jan Allbeck, and Norm Badler
October 2015

Acknowledgments

The efforts described in this volume have been made possible through the contributions of several people and funding sources. The authors wish to gratefully acknowledge their role in furthering the field of virtual crowd simulation.

Contributors. The chapters in this book include important contributions from the following researchers (in alphabetical order): Alejandro Beacco, Alexander Shoulson, Cameron Pelkey, Cory D. Boatright, Funda Durupinar, Francisco Garcia, Jennie M.Shapira, John T. Balint, Glenn Reinman, Max Gilbert, Nathan Marshak, Petros Faloutsos, Pengfei Huang, Ramon Oliva, Shawn Singh, Vivek Reddy, and Weizi Li. We also wish to thank Rita Powell, Tara Hickey, and Bria Roye for their help in preparing the bibliography for this book.

Funding Support. A portion of this research was sponsored by the Army Research Laboratory and was accomplished under Cooperative Agreement # W911NF-10-2-0016. The views and conclusions contained in this document are those of the authors and should not be interpreted as representing the official policies, either expressed or implied, of the Army Research Laboratory or the U.S. Government. The U.S. Government is authorized to reproduce and distribute reprints for government purposes notwithstanding any copyright notation herein.

Partial support for this effort is gratefully acknowledged from the U.S. Army SUBTLE MURI W911NF-07-1-0216 and the U.S. Army Night Vision and Electronic Sensor Directorate (W15P7T-06-D-E402). We would like to acknowledge the Spanish Ministry of Economy and Competitiveness and the FEDER fund, for partly supporting this work under grant number TIN2014-52211-C2-1-R. We also gratefully acknowledge Intel Corp. for their generous support through equipment and grants, as well as Autodesk, Unity, and USC ICT's SmartBody project for their support and use of software.

Mubbasir Kapadia, Nuria Pelechano, Jan Allbeck, and Norm Badler
October 2015

CHAPTER 1
Introduction

Observe any populated urban environment today. You see numerous individuals going about their everyday activities. Some will be walking to work or school, some will be waiting for buses or taxis. Others will be working on the street as police or road workers. The number and circumstances of people will vary by the time of day: differences will be apparent when businesses open or close, when noon lunchtime arrives, and when nightlife begins and ends. Under normal conditions the ebb and flow of people has some meaningful structure and is not entirely random, though at a local level the actual goals and desires of people are not readily known. Nonetheless, there is implicit structure to the movements that we see. People usually do not bump into one another. They take reasonably efficient routes toward their goals, though they may take departures from strict minimal paths in order to buy a snack, make a phone call, avoid traffic, or look into store windows. These everyday human activities present the vitality and character of an urban place through occupation, use, and traversal.

Against this rich backdrop of natural but highly varied human behavior, we seek to construct computational models of population flows and movements: crowd simulation. Computational models for moving a collection of individuals in space has a substantial history across a number of disparate fields: transportation planning for stations, airports and transfer points, evacuation analysis for emergencies in large stadiums, ships, and airplanes, and animated populated environments for movie battle scenes, virtual crowd augmentation, visual effects, and computer games. With this range of existing, useful, and economically justifiable applications, the field of crowd simulation has grown rapidly with various modeling methodologies, software tools for realization, and animation enhancements for visual fidelity.

Methodologically, a primary distinction in crowd simulation is whether the movement is represented macroscopically, much like an ambient fluid, or microscopically, in which individual agents decide, or are controlled, as to what direction they should go. The mathematical differences aside, the microscopic approach is the one we prefer, since our animated populace should look and behave as if they were motivated by often hidden internal goals and intentions. The macroscopic approach sacrifices individual movements in favor of collective measures such as flow rates and crowd density. In this volume we will only look at microscopic models: what controls move individual agents in ways that respect their personal space, their separate goals, and their internal optimizations to balance time, effort, path length, schedule, and locomotion choice.

Even this consideration of the individual at the microscopic simulation level leaves a considerable gap: knowing which way to go requires that a real person make real biomechanical choices for footsteps. That is, agents may need to move in a certain direction, at a certain speed, and with a particular orientation, and to do this they cannot merely "skate" or "slide"—they need to take steps. This component of the simulation is appropriately called "steering." A crucial link between simulation of where one wants to go and how one actually gets there requires, therefore, computational models of plausible and physiologically achievable steering behaviors. These behaviors drive footstep placement, a key factor in animation realism.

Knowing where one wants to steer must still be modulated by external factors, such as avoiding collisions with other agents and the environment. To do this in a human-like fashion requires some forethought. People do not walk, bump into things, and respond by making an alternative direction choice. This might suffice for a robotic vacuum cleaner, but it is not normal for human behavior. Direction planning requires at least local knowledge of the surrounding static and dynamic obstacles and, if they are moving, some predictions about where they might be in the next instant in order to pre-emptively avoid collisions or conflicts. These local movement decisions are crucial to the steering algorithm. The overall goal or path the agent desires will factor into the directional choices to be made at each step. For example, an agent whose path may be blocked momentarily may choose to change velocity or even stop, in preference to backing up. While any of these strategies might result in ultimate success in reaching a goal while avoiding collisions, some may appear to be unusual or even non-human choices. Again, to simulate people rather than robots, human qualities must be considered.

Discovering and quantifying these human qualities and their algorithmic realization for steering choices present many avenues for research and implementation. For example, there are procedural (algorithmic) methods to make steering choices given any possible input configuration of agents and obstacles. Others use collected data from real crowd motions to build databases of steering choices. When presented with a decision, the agent draws from the previously stored models to match as best it can its present situation, to choose and attempt an appropriate next step. By using machine learning techniques we can relax the limitations and biases of collected datasets by extending the range of realistic situations that may be encountered and handled.

Going beyond the exigencies of steering leads into larger scale planning and human activity domains. "Navigation" is the specification and determination of general ways to reach a goal. Navigation is a guide for local steering decisions, but it has a wider scope: the goal may be far away and may be subject to numerous environmental constraints, e.g., "use the sidewalks," "don't walk on the grass," "wait for the green light before crossing the street," "cross only in the crosswalk," "watch out for the wet spot," "beware of dog," etc. Navigation and steering are inseparable, so their mutual interaction must be considered for all agents in the crowd simulation. A classic example of such interaction is a contrived but informative simulation of a large number of agents on the periphery of a circle. They are required to all simultaneously walk to the point on the circle diametrically opposite their starting position as quickly as possible. Clearly each wishes to navigate along a straight line through the circle center, but just as clearly, that is not possible since each agent occupies a finite space and they can't pass through one another. Steering is necessary to achieve the desired navigation goals. Such interactions lead to novel multi-domain navigation planning algorithms.

Returning to the observer in the real world, we note that the environment has an unexpected richness of sensory information. Of course, we see the world, the agents, the obstacles, and possible collision threats, but there is also a soundscape that provides useful and even essential information to the occupants. Imagine a busy urban scene with no sound! Yet that is the basis for virtually all crowd simulation systems. The visual environment may be reproduced implicitly by having every agent instantly "know" what surrounds it, or explicitly by using a simulated visual sense to actually produce a synthetic visual image of what one could reasonably perceive. But those worlds are still silent. Implicit knowledge is a substitute for omniscient sensing and is therefore less humanly possible or realistic. Explicit visual sensing alone omits crucial auditory stimuli. To consider the importance of this omission, consider the navigation and steering issues that could be dictated solely by auditory information: the footsteps of an unseen person approaching around a blind corner, an unseen vehicle that blows its horn to warn of its impending rapid approach, a nearby whistle to get your attention, or the sound of a crash or explosion that immediately changes one's navigation goals.

Taken to another extreme, if crowd simulation agents normally possess only visual inputs, what happens if agents possess only sound inputs? This is not just an academic problem; real sight-impaired individuals may have to rely on audio cues to navigate the urban landscape. But even without impairments, normal humans use multiple senses to navigate. Sounds can be used to identify objects and activities in the environment and even used to localize them in space. This additional sensory information thus increases the human behavioral realism of the simulated agents.

Another hallmark of the human landscape is heterogeneity. People vary markedly in personality, motivations, abilities, and goals. We would expect, in all but the most rigidly controlled environments, natural variations in how different people accomplish similar goals. Part of this variation has to do with what occupation they have, whether they are in transit between places of daily living such as home, work, shopping, eating, playing, or just relaxing. Personality affects these choices and the way they are achieved. These in turn affect how the agents are animated, giving rise to perceptible differences and individualism. Such personality traits can even be associated with crowds themselves: an "angry mob," a "panicked exodus," an "attentive audience." What people remember about their world also impacts behavior and creates variations in behaviors. Two agents might require the same type of resource to achieve their current goal. Their unique previous experiences and individual recall capabilities may cause them to go in search of the required resource in different locations.

Modeling these situations requires more information than just individual traits. We need to understand the environment that the people occupy, what events are transpiring, what particular situation each agent finds itself in, and how it chooses to respond. Such choices inform navigation and behavior, but interact more with a wide set of human personality, motivation, role, and capability factors. There are two general approaches that may be taken here. One way is to imbue each agent with as many reasoning and cognitive functions as possible, mirroring or mimicking those in real people. This might be called the "Artificial Intelligence" approach to agent simulation. It is attractive because it aligns philosophically with who we believe we are. Yet it is difficult and very computationally expensive to elucidate and model human cognitive processes. Some of the computation can be eased by embedding knowledge in the virtual environment. Agents can understand how virtual objects can be used by attaching semantics to them. Alternatively, we can consider more "light-weight" agents who are recruited by and act within the confines of an externally defined "event." If the events are plausibly defined, and the recruited agents able to handle and animate the tasks handed them, the result should appear both perceptually plausible and consistent with the event characteristics.

One approach to manage and control the way agents respond to events is to bind the events themselves into a larger, more meaningful story or narrative. Looking at the urban landscape, the people have *reasons* for doing what they do. Some reasons may be mundane; some may be merely transitional so that they can get wherever they need to be. But other reasons may play as part of a larger narrative. These include job functions, such as what a road crew might do, or behaviors during more dramatic events such as a fight or a robbery. When there is a meaningful story, the situation is more engaging and realistic to the observer. Consider this: a city full of expertly animated realistic-looking humans wandering aimlessly is likely to be less engaging than a more stylized environment where interesting things are happening. Evidence for this can be found in the existence of successful video game series such as Grand Theft Auto and Assassin's Creed.

Our intention in the sequel is to follow the course just charted. While drawing heavily on previous work in crowd simulation, our exposition is designed to lead from the lowest, but crucial, footstep-steering algorithms, all the way to narrative intentions. This integrated view provides many opportunities to highlight accomplishments as well as situate and spawn new investigations. Ultimately our motivation for this effort is to produce, through computer graphics simulations rather than pure artistic means, visual experiences that possess the vibrancy, diversity, purpose, and character of real heterogeneous human crowds.

PART I

Multi-Agent Collision Avoidance

CHAPTER 2
Background

Following seminal work [225] on flocking behaviors using particle systems, the field of crowd simulation has grown into a well-developed, multi-faceted area of study [50, 271]. Steering, or goal-directed collision avoidance, is the layer of intelligence that interfaces with navigation to move an agent along its planned path by performing a series of successive local searches, taking into consideration locomotion constraints such as turning capabilities and limits on movement velocity, as well as dynamic objects in the environment such as other agents.

2.1 CENTRALIZED APPROACHES

Centralized techniques rely on a broad conformity amongst the population, such as the fluidic approach of [279]. This is an acceptable premise for efficient macroscopic group-dynamic simulations of crowd flows involving thousands of agents (e.g., stadium evacuation scenarios, urban simulations, religious pilgrimages [238], and emergency evacuations [194]). However, these approaches cannot simulate complex local agent interactions which are crucial in modeling functional, purposeful autonomous virtual humans, thus we focus on individualized, agent-based approaches. The remainder of this chapter summarizes the large body of work in agent-based steering which models each agent as an independent being having its own state and goal, performs collision avoidance with static obstacles, reacts to dynamic threats in the environment and, steers its way to its target. The interactions between agents and the environment result in the emergence of macroscopic crowd behavior such as lane formations and agents cooperating to evacuate through narrow egress points.

2.2 AGENT-BASED APPROACHES

To introduce agent individuality into a simulation, we can make steering an integral part of the agents' abilities. The simplest approach is to model each agent as a particle and to simulate interactions using basic particle dynamics [225, 226]. This approach works extremely fast, but the heuristic reactions of this technique are insufficient to create realistic human pedestrian steering behaviors.

The social force model [88, 207] simulates hypothetical forces such as repulsion, attraction, friction and dissipation for each agent. While these models take a more social interaction view of pedestrian behaviors, they are also designed to be as simple and efficient as possible. Rule-based approaches [154, 172, 193, 193] use various conditions and heuristics to identify the agent's situation, from which the rules compute a steering decision. These approaches are limited by the situations that are modeled by the rule developer and cannot generalize to handle the intractable space of all possible steering challenges.

2.2.1 DATA-DRIVEN APPROACHES

Data-driven steering tries to avoid relying on a designer's intuition and heuristics by generating local-space steering behaviors from observations of real people. In [159] video samples are compiled into a database which is queried at simulation runtime. A retrieved trajectory is assigned to an agent based solely on the similarity of the agent's surroundings to the video examples. The work of [157] uses a more constrained state space of discretized slices around an agent and focuses more on recreating group dynamics than individual steering. A similar interpersonal state space is used by [277], but in addition there is a second state space for environmental navigation consisting of a discretized view frustum. In common to all these techniques is using one collection of samples to model all navigation possibilities. Recent work [2] uses discretized pieces of real-world trajectories as the basis for navigation and manipulates these trajectories based on the possibility of future collisions. Data-driven approaches promise to capture the real essence of human crowds, but are limited to data that sufficiently captures the space of possible human interactions in high-density situations.

2.2.2 PREDICTIVE APPROACHES

The prediction and avoidance of potential dynamic collision threats results in more realistic steering behaviors, such as in the works of [63, 136, 205]. Reciprocal Velocity Obstacles [292] builds upon the concept of velocity obstacles to propose a refined method for local reactive collision avoidance using the assumption that neighboring agents will adopt similar collision-avoidance behaviors. This facilitates the efficient simulation of dense crowds without the need for explicit inter-agent communication. ClearPath [79] extends velocity obstacles to formulate collision-free navigation as a quadratic optimization problem which can be parallelized to achieve an order of magnitude performance gain.

The work in [128, 129] uses a variable-resolution egocentric representation to model agent perception and affordance for crowd simulation. This approach provides the benefit of implicit space-time planning without the overhead of modeling time as an extra dimension. Pettre and colleagues [219] use experimental data to derive a predictive model of collision avoidance in virtual walkers. Ondrej et al. [200] demonstrate a synthetic vision-based approach to predict the trajectories of neighboring agents while making steering decisions. Singh et al. [252] presents a hybrid framework that chooses between reactive, predictive, and planning-based steering policies depending on the current situation of the agent. SteerSuite [250] is an open source crowd simulator that provides developers with tools to create their own steering techniques.

2.3 LOCOMOTION SYNTHESIS

Given a steering decision, the agent is obligated to achieve movement using locomotion. Locomotion synthesis depends on how the character is being controlled. If a user controls the character with a third person, video game-like controller, it is common to work on a body root velocity basis, because the user wants to move the character responsively and in real-time so artifacts such as foot-skate can be ignored. Considering the root velocity as the input parameter can synthesize smooth, versatile and plausible locomotion animations [74, 105]. Some approaches have also used the idea of selecting animations from a Delaunay triangulation of all the available animation clips [217, 218].

Optimization-based approaches [280] are able to synthesize animations that conform to velocity and orientation constraints. However, they need a very large database and their computational time does not allow many characters in real-time. Semi-procedural animation systems [105] work with a small set of animations and use inverse kinematics only for the legs to ensure ground contact and to adapt the feet to possible slopes of the terrain, but they are unable to follow footstep trajectories.

There has been a recent surge in approaches that produce footstep trajectories for character control. They can be physically based but generated off-line [60], be generated online from an input path computed by a path planner [54], or use simplified control dynamics to produce biomechanically plausible footstep trajectories [253]. Footstep-driven animation systems [73] often produce unnatural results using purely procedural methods. The work in [34] uses a statistical dynamic model learned from motion capture data in addition to user-defined space-time constraints (such as footsteps) to solve a trajectory optimization problem. In [36], random samples of footsteps comprise a roadmap going from one point to another, used to find a minimum-cost sequence of motions matching it and then retargeted to the exact foot placements.

The work in [143, 289] performs an optimization over an extracted center of mass trajectory to maximize the physical plausibility and perceived comfort of the motion to satisfy the footprint constraints. Recent solutions [54, 287, 288] adopt a greedy nearest-neighbor approach over larger motion databases. To ensure spatial constraints, the character is properly aligned with the footsteps and reinforced with inverse kinematics, while temporal constraints are satisfied using time warping. These techniques achieve highly accurate results based on foot positioning, but their computational cost makes them unsuitable for real-time animation of large groups of agents.

2.4 CHALLENGES AND PROPOSED SOLUTIONS

We identify the main limitations in agent-based collision avoidance approaches and propose solutions that are described in detail in the subsequent chapters.

2.4.1 PARTICLE-BASED AGENT MODELS

A particle-based agent representation enables efficient state-based queries, such as collision checks, and greatly reduces the branching factor of the action space to enable efficient control. However, a particle cannot capture all the capabilities of human locomotion such as side-stepping, careful foot placement, and even nuanced upper body motions. This reduced problem domain prevents the solution of many challenging scenarios such as oncoming agents in a narrow passageway, where an agent may have to carefully step out of the way and re-orient its shoulders, allowing the other agent collision-free passage. The inverse of this is also true, where constraints imposed on a particle cannot capture bipedal locomotion constraints. A circular collision radius is a conservative estimate of the boundary of a real human and prohibits optimal packing density in crowded situations. The bounding volume of real humans constantly changes and deforms while maneuvering around different obstacle and agent configurations.

We use a locomotion model that represents a character's stepping state and footstep actions using a 2D approximation of an inverted spherical pendulum. Our model encapsulates the energy cost of footstep actions, and can be easily integrated with an efficient short-horizon planner to generate a timed sequence of footsteps that existing motion synthesis techniques can follow exactly. During planning, our locomotion model not only constrains characters to navigate with realistic locomotion, but more importantly, it also enables characters to control finer-grain subtle *navigation* behaviors that are possible with exact footsteps. Our approach can navigate crowds of hundreds of individual characters with collision-free, natural behaviors in real-time. Chapter 3 describes this approach.

2.4.2 DECOUPLING BETWEEN STEERING AND LOCOMOTION

There exists a uni-directional communication interface between steering and locomotion where steering outputs only a force or velocity vector to an animation system, without necessarily conforming to the constraints and capabilities of human-like movement. Steering trajectories may have discontinuous velocities, oscillations, awkward orientations, or may try to move a character unnaturally. A vector-based steering interface does not have enough information to indicate certain maneuvers such as side-stepping versus reorienting the torso, stepping backward versus turning around, planting a foot to change momentum quickly, or carefully placing steps in exact locations.

This simplistic approach lacks control over how a character should truly animate and often results in visual artifacts when an animated virtual human follows a trajectory output by steering. The locomotion system is not always able to accurately follow the steering trajectory, causing deviations and possibly even collisions, and conversely the steering layer may not always be aware of the abilities of the locomotion system. To offset these limitations, we present an animation system that can accurately follow footstep trajectories that are generated by a footstep-based collision avoidance approach. Chapter 3 describes this approach in detail.

2.4.3 GENERALIZATION AND APPLICABILITY OF DATA-DRIVEN APPROACHES

As simulated environments grow in complexity and interactivity, it quickly becomes intractable to predict *a priori* the possible scenarios an agent will encounter. There are two main types of data-driven crowd simulation, trained models [22] and database queries [17,37]. Individually, these approaches suffer from limitations. First, trained models fit a single, monolithic model which must generalize over all the training data, requiring accuracy in some scenarios to be sacrificed for others. Databases require unwieldy amounts of data which become impractical to acquire, store and search. As these approaches use trajectories of real humans as training data, there is a logistical challenge for data acquisition and state coverage. To address these limitations we combine data-driven solutions with machine learning to obtain an algorithm that can handle novel situations. By using comprehensive synthetic training data to learn steering policies for different subsets of scenarios, results generalize well across the space of all possible agent interactions while providing a solution that is scalable, yet efficient. Chapter 5 describes our approach in detail.

CHAPTER 3

Footstep-based Navigation and Animation for Crowds

Shawn Singh, Mubasir Kappadia, Glenn Reinman, and Petros Faloutsos

3.1 INTRODUCTION

The important ability of autonomous virtual characters to locomote from one place to another in dynamic environments has generally been addressed as the combination of two separate problems: (a) navigation and (b) motion synthesis. For the most part, research along these two directions has progressed independently of one another. The vast majority of navigation algorithms use a trivial locomotion model, outputting a simple force or velocity vector and assuming that motion synthesis can realistically animate a character to follow these arbitrary commands. This simplistic interface between navigation and motion synthesis introduces important limitations:

- *Limited locomotion constraints:* Very few navigation algorithms account for locomotion constraints. Trajectories may have discontinuous velocities, oscillations, awkward orientations, or may try to move a character during the wrong animation state, and these side-effects make it harder to animate the character intelligently. For example, a character cannot easily shift momentum to the right when stepping with the right foot, and a character would rarely side-step for more than two steps at a steady walking speed.

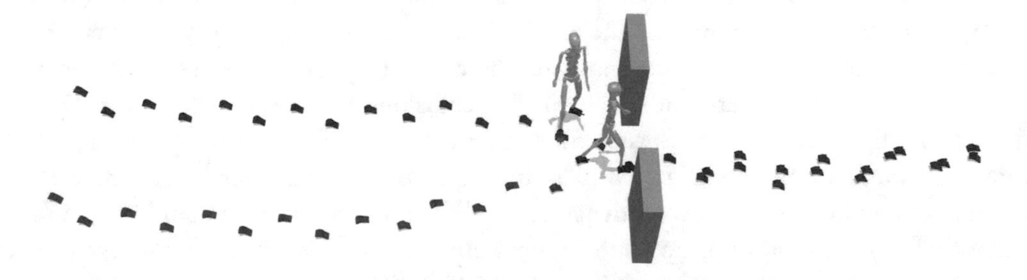

Figure 3.1: Autonomous characters navigating with our footstep locomotion model. Footstep navigation allows characters to maneuver with spatial precision, temporal precision, and natural locomotion.

- *Limited navigation control:* Existing navigation systems usually assume that motion synthesis will automatically know when to perform subtle maneuvers, such as side-stepping versus reorienting the torso, stepping backward versus turning around, stopping and starting, planting a foot to change momentum quickly, or carefully placing footsteps in exact locations. However, there are many cases where navigation should be aware of such options and have some control over their execution. For example, a character standing next to a wall would probably turn around toward the open space, not towards the wall. A character could better navigate through tight spaces, including dynamic spaces between other navigating characters, if its collision radius were adaptable.

Existing systems rely on robust motion synthesis to address such problems, but this forces motion synthesis to make undesirable trade-offs: either the motions follow navigation commands precisely, which can cause discontinuous or awkward animations, or the motions ensure proper, natural locomotion regardless of the requested timing, which can cause delayed reactions and collisions. In either case, the navigation intelligence of the character is diminished in the final result that a viewer sees. For example, many high quality motion synthesis techniques based on motion graphs have a latency between receiving and executing a command, because of the time it takes to continue animating until a valid transition is possible. These issues motivate the need for navigation to be more locomotion-aware and the need for better integration between navigation and motion synthesis.

In this chapter, we propose one possible solution to this problem: generating sequences of footsteps as the interface between navigation and motion synthesis. Foot placement and timing is an intuitive and representative abstraction for most locomotion tasks. Footsteps offer finer control which allows navigation to choose more natural and nuanced actions. At the same time, the series of footsteps produced by navigation communicates precise, unambiguous spatial and timing information to motion synthesis. Existing motion synthesis algorithms can animate a character to follow these generated and timed footsteps exactly [38, 39, 47, 72, 142, 276, 289].

We use a space-time planning approach to dynamically generate a short-horizon sequence of footsteps. Footsteps are represented as a 2D locomotion trajectory that approximates an inverted spherical pendulum model. The pivot of the pendulum trajectory represents a foot's location and orientation. The character navigates by computing an efficient sequence of space-time trajectories (footsteps) that avoids time-varying collisions, satisfies footstep constraints for natural locomotion, and minimizes the effort to reach a goal. In its most general form, this is a difficult nonholonomic optimization planning problem in continuous space where the configuration and action spaces must be dynamically re-computed for every plan [156]. We mitigate these challenges by: (a) using an *analytic* approximation of inverted pendulum dynamics, (b) allowing near-optimal plans, (c) exploiting domain-specific knowledge, and (d) searching only a limited horizon of footsteps; our approach generates natural footsteps in real-time for hundreds of characters. Characters successfully avoid collisions with each other and produce dynamically stable footsteps that correspond to natural and fluid motion, with precise timing constraints. Because the most significant biomechanics and timing constraints are already taken into account in the navigation locomotion model, using a motion synthesis algorithm that follows footsteps is straightforward and results in animations that are nuanced, collision-free, and plausible. The relevant publications pertaining to this chapter can be found here [15, 253, 254].

3.2 LOCOMOTION MODEL

The primary data structure in our locomotion model is a *footstep*, storing: (1) the trajectory of the character's center of mass, including position, velocity, and timing information, (2) the location and orientation of the foot itself, and (3) the cost of taking the step. We next describe how the model computes these aspects of a footstep, as well as the constraints associated with creating a footstep.

3.2.1 INVERTED PENDULUM MODEL

Our approach is inspired by the *linear inverted pendulum* model [109] which produces equations of motion derived by constraining the center of mass to a horizontal 2D plane. The passive dynamics of a character's center of mass pivoting over a footstep located at the origin is then described by:

$$\ddot{x}(t) = \frac{g}{l}x, \tag{3.1}$$

$$\ddot{y}(t) = \frac{g}{l}y, \tag{3.2}$$

where g is the positive gravity constant, l is the length of the character's leg from the ground to the center of mass (i.e., the length of the pendulum shaft), and (x, y) is the time-dependent 2D position of the character's center of mass (i.e., the pendulum weight). Note that both x and y have the same dynamics as the basic planar inverted pendulum when using the small-angle approximation, $x \simeq \sin x$. This is valid for the region of interest since the angle between a human leg and the vertical axis rarely exceeds 30 degrees.

Here we further assume that l remains constant—a reasonable approximation within the small-angle region. Then, given initial position (x_0, y_0) and initial velocity (v_{x_0}, v_{y_0}), the solution for $x(t)$ is a general hyperbola:

$$x(t) = \frac{(kx_0 + v_{x_0})e^{kt} + (kx_0 - v_{x_0})e^{-kt}}{2k}, \tag{3.3}$$

and $y(t)$ has a similar solution, where $k = \sqrt{\frac{g}{l}}$.

We use the second-order Taylor series approximation of this parabola:

$$x(t) = x_0 + v_{x_0}t + \frac{x_0k^2}{2}t^2, \tag{3.4}$$

$$y(t) = y_0 + v_{y_0}t + \frac{y_0k^2}{2}t^2. \tag{3.5}$$

Finally, this parabola is translated, rotated, reflected, and re-parameterized from world space into a canonical parabola in local space:

$$x(t) = v_{x0} t, \tag{3.6}$$
$$y(t) = \alpha t^2, \tag{3.7}$$
$$\dot{x}(t) = v_{x0}, \tag{3.8}$$
$$\dot{y}(t) = 2\alpha t, \tag{3.9}$$

such that both v_{x_0} and α are positive.

Evaluating the trajectory. Equations 3.6–3.9 allow us to *analytically* evaluate the position and velocity of a character's center of mass at any time. The implementation is extremely fast: first, the local parabola and its derivative are evaluated at the desired time, and then the local position and velocity are transformed from the canonical parabola space into world space. This makes it practical to search through many possible trajectories for many characters in real-time.

3.2.2 FOOTSTEP ACTIONS

The state of the character $s \in \mathbb{S}$ is defined as follows:

$$s = \langle (x, y), (\dot{x}, \dot{y}), (f_x, f_y), f_\phi, I \in \{\texttt{left},\texttt{right}\} \rangle, \tag{3.10}$$

where (x, y) and $(\dot{x}, \dot{y})$ are the position and velocity of the center of mass of the character at the *end* of the step, (f_x, f_y) and f_ϕ are the location and orientation of the foot, and I is an indicator of which foot (left or right) is taking the step. The state space $\mathbb{S}$ is the set of valid states that satisfy the constraints of the locomotion model described below.

A *footstep action* transitions a character from one state to another. An action $a \in \mathbb{A}$ is given by:

$$a = \langle \phi, v_{\text{desired}}, T \rangle, \tag{3.11}$$

where ϕ is the desired orientation of the parabola, v_{desired} is the desired initial speed of the center of mass, and T is the desired time duration of the step. The action space $\mathbb{A}$ is the set of valid footstep actions, where the input and output states are both valid.

A key aspect of the model is the transition function, s' = **createFootstep**(s, a). This function receives a desired footstep action a and a state s and returns a new state s' if the action is valid. It is implemented as follows. First, ϕ, which indicates the orientation of the parabola, is used to compute a transform from world space to local parabola space. Then, the direction of velocity $(\dot{x}, \dot{y})$ from the end of the previous step is transformed into local space, normalized, and re-scaled by the desired speed v_{desired}. With this local desired velocity, there is enough information to solve for α, and then Equations 3.6–3.9 are used to compute (x, y) and $(\dot{x}, \dot{y})$ at the end of the next step. The foot location is always located at $(f_x, f_y) = (0, -d)$ in local space, where d is a user-defined parameter that describes the distance between a character's foot and center at rest. Additionally, for every footstep, an interval of valid foot orientations is maintained, and the exact foot orientation within this interval is chosen with a fast, simple post-process. (We will discuss foot orientations below.) Finally, all state information is transformed back into world space, which serves as the input to create the next footstep. During the entire process, checks are made to verify that the transition and new state are valid based on constraints, described next.

3.2.3 LOCOMOTION CONSTRAINTS

Footstep orientation. Intuitively, it would seem that footstep orientations must be an additional control parameter when creating a footstep. However, the choice of footstep orientation seems to have no direct effect on the dynamics of the center of mass trajectory; in our experience the foot orientation only constrains the options for future trajectories and future footsteps. To implement this constraint, we compute an interval $[f\phi_{\text{inner}}, f\phi_{\text{outer}}]_{next}$ of valid foot orientations when creating a footstep. This interval is constrained by the same interval from the previous step, and further constrained by the choice of parabola orientation ϕ used to create the next footstep:

$$[f\phi_{\text{inner}}, f\phi_{\text{outer}}]_{next} = [f\phi_{\text{outer}}, f\phi_{\text{inner}} + \frac{\pi}{2}] \cup [\phi, (\dot{x}, \dot{y})]. \tag{3.12}$$

In words, the constraining interval $[\phi, (\dot{x}, \dot{y})]$ describes that the character would not choose a foot orientation that puts its center of mass on the outer side of the foot. The ordering of bounds in these intervals insures that the next foot's outer bound is constrained by the previous foot's inner bound. Similarly, a human would rarely orient its next footstep more outward than the orientation of their momentum—this constraint is unintuitive because it is easy to overlook momentum. Finally, parameters can be defined that allow a user to adjust these constraints, for example, to model a slow character that has difficulty with sharp, quick turns.

Space-time collision model. For any given footstep, this model can compute the *time-varying* collision bounds of the character at any exact time during the step. Thus, to determine if a footstep may cause a collision, we iterate over several time-steps within the footstep and query the collision bounds of other nearby characters for that time. The collision bounds are estimated using 5 circles, as depicted in Fig. 3.2(c). If a potential footstep action causes any collision, that footstep is considered invalid. For additional realism, a character can only test collisions with other characters that overlap its visual field.

User parameters. Our locomotion model offers a number of intuitive parameters that a user can modify. These parameters include the preferred value and limits of the duration of the step, the preferred value and limits on the stride length, the desired velocity of the center of mass and its limits, the interval of admissible foot orientations described above and its bounds, etc. By modifying these parameters a user can practically create new locomotion models. For example, restricting the range of some of the parameters results in a locomotion model that may reflect motion limits associated with the elderly.

Our model uses common default values for these parameters. For example, the desired velocity is set based on the Froude ratio heuristic, Fr. For a typical human-like walking gait,

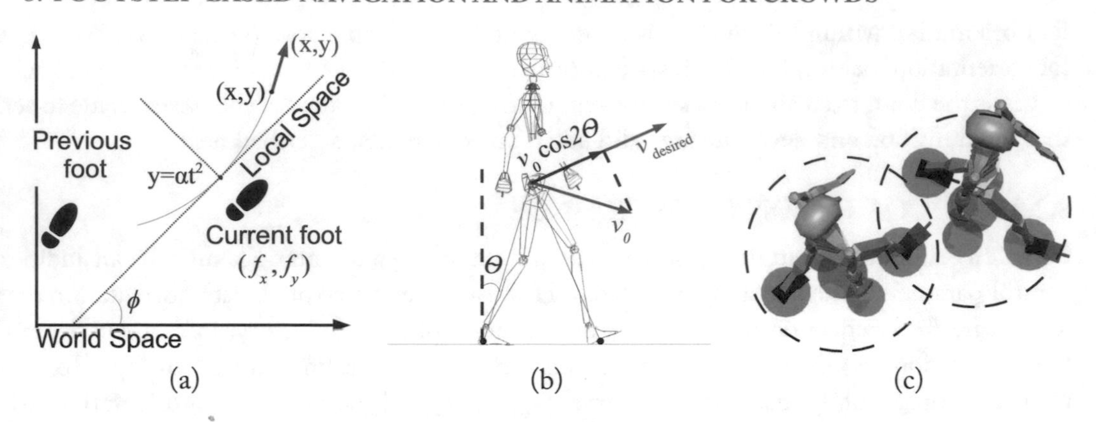

Figure 3.2: Our footstep locomotion model. **(a)** Depiction of state and footstep action parameters. The state includes the 2D approximation of an inverted spherical pendulum trajectory. **(b)** A sagittal view of the pendulum model used to estimate energy costs. **(c)** The collision model uses 5 circles that track the torso and feet over time, allowing tighter configurations than a single coarse radius.

an average value of the Froude ratio is 0.25 [19]. The associated desired velocity is $v = \sqrt{glFr}$, where g is the acceleration of gravity and l is the height of the pelvis.

3.2.4 COST FUNCTION

We define the cost of a given step as the energy spent to execute a footstep action. We model three forms of energy expenditure for a step: (1) ΔE_1, a fixed rate of energy the character spends per unit time, (2) ΔE_2, the work spent due to ground reaction forces to achieve the desired speed, and (3) ΔE_3, the work spent due to ground reaction forces accelerating the center of mass along the trajectory. The total cost of a a footstep action transitioning a character from s to s' is given by:

$$c(s, s') = \Delta E_1 + \Delta E_2 + \Delta E_3. \tag{3.13}$$

Fixed energy rate. The user defines a fixed rate of energy spent per second, denoted as P. For each step, this energy rate is multiplied by the time duration of the step T to compute the cost:

$$\Delta E_1 = P \cdot T. \tag{3.14}$$

This cost is proportional to the amount of time it takes to reach the goal: thus minimizing this cost corresponds to the character trying to minimize the time it spends walking to its goal. We found that good values for P are roughly proportional to the character's mass.

Ground reaction forces. We model three aspects of ground reaction forces that are exerted on the character's center of mass, based on the biomechanics literature [148]. The geometry and notation of the cost model is shown in Figure 3.2. First, at the beginning of a new step, some of the character's momentum dissipates into the ground. We estimate this dissipation with straightforward trigonometry, reducing the character's speed from v_0 to $v_0 \cos(2\theta)$. Second, in order to resume a desired speed, the character actively exerts additional work on its center of mass, computed as:

$$\Delta E_2 = \frac{m}{2} \left| (v_{\text{desired}})^2 - (v_0 \cos(2\theta))^2 \right|. \tag{3.15}$$

This cost measures the effort required to choose a certain speed. At every step, some energy is dissipated into the ground, and if a character wants to maintain a certain speed, it must actively add the same amount of energy back into the system. On the other hand, not all energy dissipates from the system after a step, so if the character wants to come to an immediate stop, the character also requires work to remove energy from the system. Minimizing this cost corresponds to finding footsteps that require less effort, and thus tend to look more natural. Furthermore, when walking with excessively large steps, $\cos(2\theta)$ becomes smaller, implying that more energy is lost per step.

There is much more complexity to real bipedal locomotion than this cost model; for example, the appropriate bending of knees and ankles, and the elasticity of human joints can significantly reduce the amount of energy lost per step, thus reducing the required work of a real human versus this inverted pendulum model. However, while the model is not an accurate measurement, it is quite adequate for *comparing* the effort of different steps.

ΔE_2 captures only the cost of changing a character's momentum at the beginning of each step, but momentum may also change during the trajectory. For relatively straight trajectories, this change in momentum is mostly due to the passive inverted pendulum dynamics that requires no active work. However, for trajectories of high curvature, a character spends additional energy to change its momentum. We model this cost as the work required to apply the force over the length of the step, weighted by constant w:

$$\Delta E_3 = w \cdot F \cdot \text{length} = w \cdot m\alpha \cdot \text{length}. \tag{3.16}$$

Note that α is the same coefficient in Eq. 3.7: the acceleration of the trajectory. α increases if the curvature of the parabola is larger, and also if the speed of the character along the trajectory is larger. Therefore, minimizing this cost corresponds to preferring straight steps when possible, and preferring to go slower (and consequently, take smaller steps) when changing the direction of momentum significantly. The weight w can be adjusted to prioritize whether the character prefers to stop (it costs less energy to avoid turning) or walk around an obstacle (it costs more energy to stop). In general we found good values of weight w to be between 0.2 and 0.5. The interpretation is that twenty to fifty percent of the curvature is due to the character's effort.

3.3 PLANNING ALGORITHM

Our autonomous characters navigate by planning a sequence of footsteps using the locomotion model described above. In this section we detail the implementation and integration of the short-horizon planner that outputs footsteps for navigation.

Generating discretized footstep actions. The choices for a character's next step are generated by discretizing the action space $\mathbb{A}$ in all three dimensions and using the **createFootstep**(s, a) function to compute the new state and cost of each action. We have found that v_{desired} and T can be discretized extremely coarsely, as long as there are at least a few different speeds and timings. Most of the complexity of the action space lies in the choices for the parabola orientation, ϕ. The choices for ϕ are defined relative to the velocity $(\dot{x}, \dot{y})$ at the end of the previous footstep, and the discretization ranges from almost straight to almost U-turns. The first choice that humans would usually consider is to step directly towards the local goal. To address this, we create a special option for ϕ that would orient the character directly towards its goal. With this specialized goal-dependent option, we found it was possible to give fewer fixed options for ϕ, focusing on larger turns. Without this option, even with a large variety of choices for ϕ, the character appears to steer towards an offset of the actual goal and then takes an unnatural corrective step.

Short-horizon best-first search. Our planner uses a short-horizon best-first search algorithm that returns a near-optimal path, $\mathtt{P}^s_k = \{s_{start}, s_1, s_2 ... s_k\}$, from a start state s_{start} towards a goal state s_{goal}.

The cost of transitioning from one state to another is given by $\mathtt{c}(s, s')$, described by Equations 3.13–3.16. The heuristic function, $h(s)$ estimates the energy along the minimal path from s to s_{goal}:

$$h(s) = \mathtt{c}_{min} \times \min |\mathtt{P}^c_g|, \tag{3.17}$$

where $\mathtt{c}_{min}$ is the minimum energy spent in taking one normal footstep action, and $\min |\mathtt{P}^c_g|$ is the number of steps along the shortest distance to the goal.

The path returned is complete, i.e., $s_k = s_{goal}$ if s_{goal} lies within the horizon of the planner. The horizon of the planner, $\mathtt{N}_{max}$ is the maximum number of nodes that can be expanded by the planner for a single search. As $\mathtt{N}_{max} \rightarrow \infty$, the path returned by the planner is complete. For efficiency reasons, however, we limit the value of $\mathtt{N}_{max}$ to reasonable bounds. If the planner reaches the maximum limit of nodes to be expanded without reaching the goal, it instead constructs a path to a state from the closed list that had the best heuristic. Intuitively, this means that if no path is found within the search horizon, the planner returns a path to the node that has the most promise of reaching the goal.

User parameters. Higher level steering decisions can be modeled as hard constraints or soft constraints on the planner. Hard constraints can essentially prune the search space, while soft constraints affect the cost of making certain decisions. For example, in a city landscape, sidewalks can be considered less costly to traverse than roads.

3.4 EVALUATION

Performance and search statistics are shown in Table 3.1. One reason that our planner is fast is because of the scope of footsteps: a short horizon plan of 5–10 footsteps takes several seconds to execute but only a few milliseconds to compute. The amortized cost of updating a character at 20Hz is shown in the table.

Table 3.1: Performance of our footstep planner for a character. The typical worst case plan generated up to 5,000 nodes

	Egress 200 agents	2-way traffic 200 agents	700 obstacles 500 agents
Avg. # nodes generated	137	234	261
Avg. # nodes expanded	82	190	192
Planner performance	1.6 ms	4.4 ms	3 ms
Amortized cost 20 Hz	0.037 ms	0.1 ms	0.11 ms

The space-time aspect of our planner helps the character to exhibit predictive, cooperative behaviors. It can solve challenging situations such as potential deadlocks in narrow spaces; e.g., Figures 3.3 and 3.4 depict a character side stepping to allow the other pedestrian to pass first. Because the doorway is barely wide enough to fit a single pedestrian, many other navigation techniques would rely on collision prevention at the walls until the character eventually finds the doorway.

Figure 3.3: A character navigates through oncoming traffic at a doorway.

Figure 3.4: A character side-steps and yields to the other pedestrian, and then precisely navigates through the narrow doorway.

Our time-varying collision model, as shown in Figure 3.2, dynamically adjusts the bounds of a character more realistically than fixed size disks, and thus allows much tighter spacing in crowded conditions. In Figure 3.5 a group of characters tightly squeeze through a narrow door. The planner and locomotion model offer a number of intuitive and fairly detailed parameters to interactively modify agent behaviors. In Figure 3.6, restricting the range of footstep orientations results in additional turning steps.

Figure 3.5: An egress simulation. Characters do not get stuck around the corners of the glass door.

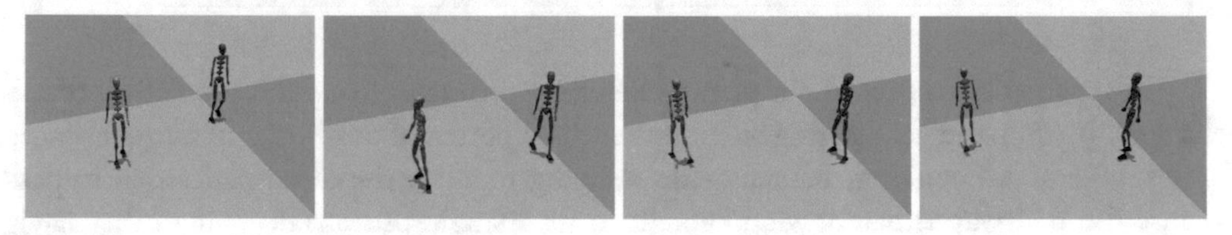

Figure 3.6: Two characters performing a U-turn, with different user-defined ranges of foot orientations. The character on the right needs more steps to turn.

3.4.1 INTERFACING WITH MOTION SYNTHESIS

Interfacing with this footstep navigation method is very flexible. It can output properly timed footsteps, center of mass trajectories and velocities at any given time. Of course, to take advantage of the full intelligence of our approach, the motion synthesis should ideally be able to follow properly timed sequences of footsteps, including footstep orientation. For offline, high-quality animation systems, our method can be modified to output a more accurate center of mass trajectories by using a hyperbola described in Equation 3.3 instead of the parabolic approximation, and by modifying the locomotion model and planner accordingly. The hyperbolic form is slower to evaluate, but it and its derivatives are still analytic, and the center of mass trajectories more accurately represent the inverted pendulum dynamics—in particular, the way a human's center of mass "lingers" at the apex of the motion.

CHAPTER 4

Following Footstep Trajectories in Real Time

Alejandro Beacco, Nuria Pelechano, Mubbasir Kapadia, and Norman I. Badler

4.1 ANIMATING FROM FOOTSTEPS

Virtual human navigation schemes that generate center of mass (COM) trajectories [87, 207, 225, 279] lead to foot-sliding artifacts or awkward motions when the root orientation and the displacement vector of the animation do not match. Different animations can be blended by tweaking some of the upper body joints [210] to minimize artifacts, at the expense of constant updates to account for the decoupling between the navigation algorithm and the animation system. To alleviate these problems we just examined footstep-based control systems [54, 253] that output a sequence of space-time foot-plants.

Our goal now is to synthesize human-like animations that accurately follow the footsteps. This is sometimes called the *stepping stone* problem. We present an online animation synthesis technique for fully embodied virtual humans that satisfies foot placement constraints for a large variety of locomotion speeds and styles (see Fig. 4.1). Moreover, the output trajectory can be modified by external perturbations such as uneven terrain.

Given a database of motion clips, we precompute multiple parametric spaces based on the motion of the root and the feet. A root parametric space is used to compute a weight for each available animation based on root velocity. Parametric spaces for each foot are based on a Delaunay

Figure 4.1: An autonomous virtual human navigating a challenging obstacle course (a), walking over a slope (b), exercising careful foot placement constraints including side-stepping (c), speed variations (d), and stepping back (e). The system can handle multiple agents in real time (f).

triangulation of the graph of possible foot landing positions. For each foot parametric space, blending weights are calculated as the barycentric coordinates of the next footstep position for the triangle in the graph that contains it. These weights are used for synthesizing animations that accurately follow the footstep trajectory while respecting the singularities of the different walking styles captured in the database. Blending weights calculated as barycentric coordinates are used to reach the desired foot landing by interpolating between several proximal animations, and inverse kinematics is used to adjust the final position of the support foot to correct for minor offsets, foot step orientation and the angle of the underlying floor. Compared to [105], we exploit the combination of multiple parameter spaces for footstep-precision control. This reduces the dimensionality of the problem, compared to [288].

Since foot parametric space only considers final landing positions of the feet without taking into account root velocity, animations selected might satisfy position constraints but introduce discontinuities in root velocity. To incorporate root velocity fidelity we present a method that can integrate both foot positioning and root velocity. Our method also allows the system to recover nicely when the input foot trajectory contains steps that are impossible to perform with the given set of animations (for example, due to extreme distance between steps). The user also can control over the trade-off between footstep accuracy and root velocity. Footstep trajectories can be applied to different walking styles and even running motions with a short foot flight phase.

The method is evaluated on a variety of test cases and error measurements provide a quantitative analysis. Our current implementation can efficiently generate visually appealing animations for over sixty agents in real time (25 FPS), and over a hundred characters at 13 FPS, without compromising motion fidelity or character control, and can be easily integrated into existing crowd simulation schemes.

4.2 FRAMEWORK OVERVIEW

Requirements for real-time agent animations depend on the application. In some, the user may only need to control the direction of movement and speed of the root, but there are other situations where finer control of foot positioning is necessary. For example, the user may want to respect different walking gaits depending on the terrain, make the character step over stones to cross a river, or walk through some space full of holes but avoid falling. Therefore, our framework animates agents following footstep trajectories, while still subject to guidance from the COM when necessary.

We define two parametric spaces based on the position of each foot, Ω_{f_L} and Ω_{f_R}, and switch between the two depending on the swing foot, and also a parametric space based on the root movement Ω_r. Our technique takes into account both displacement (from Ω_{f_L} and Ω_{f_R}) and speed (from Ω_r) to ensure the satisfaction of both spatial and temporal constraints. The system (Fig. 4.2) provides the user with the flexibility to choose between different control granularities ranging from exact foot positioning to exact root velocity trajectories.

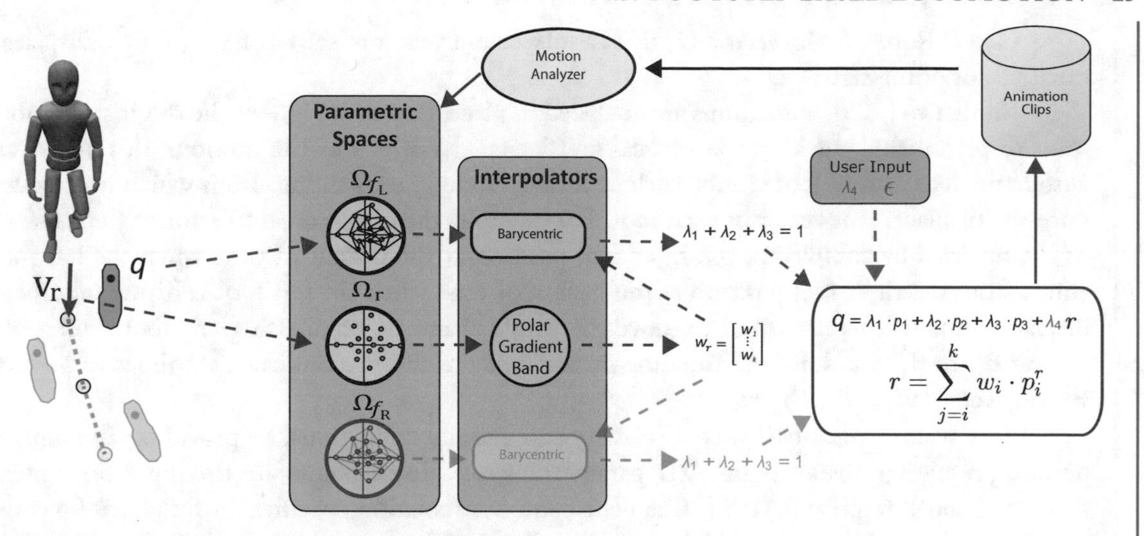

Figure 4.2: Online selection of the blend weights to accurately follow a footstep trajectory. Ω_r uses a gradient band polar based interpolator [105] to give a set of weights w_j, which are then used by the barycentric coordinates interpolator to trade-off between footstep and COM accuracy.

4.3 FOOTSTEP-BASED LOCOMOTION

The main goal of the Footstep-based Locomotion Controller is to accurately follow a footstep trajectory, i.e., to animate a fully articulated virtual human to step onto a series of footplants with space and velocity constraints. The system must meet real-time constraints for a group of characters, should be robust enough to handle sparse motion clips, and needs to produce synthesized results that are devoid of artifacts such as foot sliding and collisions.

4.3.1 MOTION CLIP ANALYSIS

We begin with a collection of cyclic motion clips. Although cyclic animations are not strictly required, they do assist in defining smoother transitions between consecutive footsteps and are preferred by most standard animation systems [105]. Individual footsteps must be extracted from the clips; each cycle contains two steps, one starting with the left foot on the floor, and one starting with the right foot. A step is defined as the action where one foot of the character starts to lift-off the ground, moves in the air and finishes when it is again planted on the floor. We say that a footstep corresponds to a particular foot when that foot is the one performing the step action. The other foot that stays in contact with the floor is called the supporting foot, since it supports the weight of the body.

In an offline analysis, each motion clip m_i is annotated with the following information:

(1) $\mathbf{v}_i^r$: Root velocity vector. (2) $\mathbf{d}_i^L$: Displacement vector of the left foot. (3) $\mathbf{d}_i^R$: Displacement vector of the right foot.

Similar to [105], animations are analyzed in place, that is, we ignore the original root forward displacement, but keep the vertical and lateral deviations of the motion. This allows an automatic detection of foot events, such as lifting, landing, or planting, from which we can deduce the displacement vector of each foot. For example, the displacement vector of the left foot $\mathbf{d}_i^L$ is obtained by subtracting the right foot position at the instant of time when the left foot lands, from the right foot position at the instant of time when the left foot is lifting off. These displacements will be later used to move the whole character, eliminating any foot sliding. By adding $\mathbf{d}_i^L$ to $\mathbf{d}_i^R$ and knowing the time duration of the clip, we can calculate the average root velocity vector $\mathbf{v}_i^r$ of the clip m_i.

This average velocity is used to classify and identify animations, by providing an example point representing a velocity in a 2D parametric space. These points are the input for a polar gradient band interpolator [105]. Gradient band interpolation specifies an influence function associated with each example, which creates gradient bands between the example point and each of the other example points. These influence functions are normalized to get the weight functions associated with each example. However the standard gradient band interpolation is not well suited for interpolation of examples based on velocities. The polar gradient band interpolation method gives better weight functions over example points that represent velocities, since it accommodates differences in direction and magnitude rather than differences in the Cartesian vector coordinate components.

Each motion clip is split into two separate step animations A_i^L for the left foot and A_i^R for the right foot. For each foot, we need to calculate all the possible positions that can be reached based on the set of animation steps available. Since the same analysis is performed for each foot independently, we will not need to differentiate between left and right in the exposition. For each individual step animation A_i and given an initial root position, we want to extract the foot landing position p_i, if the corresponding section of its original clip was played. This is calculated by summing the root displacement during the section of the animation with the distance vector between the root projection over the floor and the foot position in the last frame.

The set $\{p_i | \forall i \in [1, n]\}$ where n is the number of step animations, provides a point cloud. Figure 4.3 shows the Delaunay triangulation that is calculated for the point cloud of landing positions. This triangulation is queried in real time to determine the simplex that contains the next footstep in the input trajectory. Once the triangle is selected, we will use its three vertices p_1, p_2 and p_3 to compute the blending weights for each of the corresponding animations A_1, A_2, and A_3.

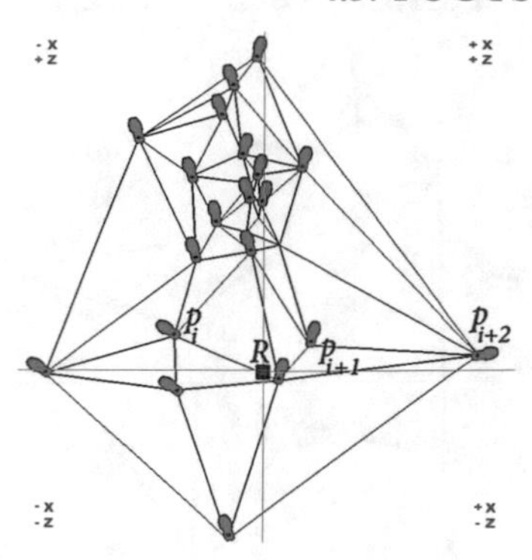

Figure 4.3: Delaunay triangulation for the vertices representing the landing positions (p_i, p_{i+1}, p_{i+2},...) of the left foot when the root, R is kept in place.

4.3.2 FOOTSTEP AND ROOT TRAJECTORIES

Our system can work with both footstep trajectories and COM trajectories. A footstep trajectory will be given as an ordered list of space-time positions with orientations, whether it is precomputed or generated on-the-fly.

The input footstep trajectory may be accompanied by its associated root trajectory (a space-time curve, rather than a list of points, and an orientation curve), or else we can automatically compute it from the input footsteps by interpolation. This is achieved by computing the projection of the root on the ground plane, as the midpoint of the line segment joining two consecutive footsteps. The root orientation is then computed as the average between the orientation vectors of each set of consecutive steps. This sequence of root positions and orientations can be interpolated to approximate the motion of the root over the course of the footstep trajectory.

4.3.3 ONLINE SELECTION

During runtime, the system animates the character towards the current target footstep. If the target is reached, the next footstep along the trajectory is chosen as the next target. For each footstep q_j in the input trajectory $\{q_1, q_2, q_3, ..., q_m\}$ we need to align the Delaunay triangulation graph with the current root position and orientation. Then the triangle containing the next foot position is selected as the best match to calculate the weights required to nicely blend between the three animations in order to achieve a footstep that will land as close as possible to the desired destination position q_j (Fig. 4.4). Notice that these weights are applied equally to all the joints in the skeleton, which means that at this stage we cannot accurately adjust the specific foot orientation required by each footstep in the input trajectory.

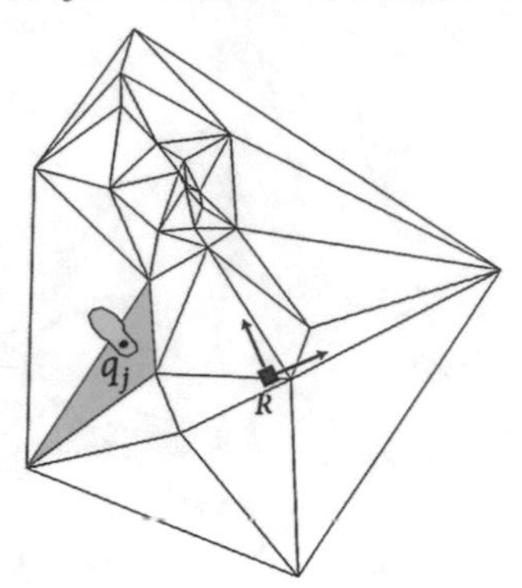

Figure 4.4: By matching root position and orientation, we can determine the triangle containing the destination position for the landing position q_j.

4.3.4 INTERPOLATION

Footstep parameters change between successive footplants, remaining constant during the course of a single footstep (several frames of motion). Therefore we need to compute the best interpolation for each footstep, blend smoothly between consecutive steps, and apply the right transformation to the root in order to avoid foot-sliding or intersections with the ground.

To meet these requirements, we use a barycentric coordinate-based interpolator in Ω_{f_L} and Ω_{f_R}, and constrain the solution based on the weights computed in Ω_r. This allows us to animate a character at the granularity of footsteps, while simultaneously accounting for the global COM motion of the full body.

If we only consider the footstep parametric space, then the vertices of the selected triangle are those that can provide the best match for the desired foot position. The barycentric coordinates of the desired footstep are calculated for the selected triangle as the coordinates that satisfy:

$$q_j = \lambda_1 \cdot p_1 + \lambda_2 \cdot p_2 + \lambda_3 \cdot p_3, \qquad \lambda_1 + \lambda_2 + \lambda_3 = 1 \tag{4.1}$$

where p_1, p_2, and p_3 are the positions of the foot landing if we run the step animations A_1, A_2, and A_3, respectively. The calculated barycentric coordinates are then used as weights for the blending between animations. A nice property of the barycentric coordinates is that the sum equals 1, which is a requirement for our blending. Finally in order to move the character towards the next position, we need to displace the root of the character adequately to avoid foot sliding. The final root displacement vector, $\mathbf{d}_j^r$ is calculated as the weighed sum of the root's displacement of the three selected animation steps (Eq. 4.2), and changes in orientation of the input root trajectory are applied as rotations over the ball of the supporting foot.

$$\mathbf{d}_j^r = \lambda_1 \cdot \mathbf{d}_1^r + \lambda_2 \cdot \mathbf{d}_2^r + \lambda_3 \cdot \mathbf{d}_3^r \tag{4.2}$$

This gives a final root displacement that is the result of interpolating between the three root displacements in order to avoid any foot sliding. The barycentric coordinates also provide the linear interpolation required between three points in 2D space to obtain the position q_j. This is an approximation to the real landing position that the character will reach, as the result of blending the different poses of the three animation clips, using spherical linear interpolation (SLERP) with a simple iterative approach as described in [245].

Therefore there will be an offset between the desired position q_j and the position reached after interpolating the three animations. To illustrate this offset, Fig. 4.5 shows the points sampled to compute barycentric coordinates in black, and in blue the real landing positions achieved after applying the barycentric weights to the animation engine and performing blending using SLERP. In order to correct this small offset at the same time that we adjust the feet to the elevation of the terrain and orient the footstep correctly, we incorporate a fast and simple IK solver.

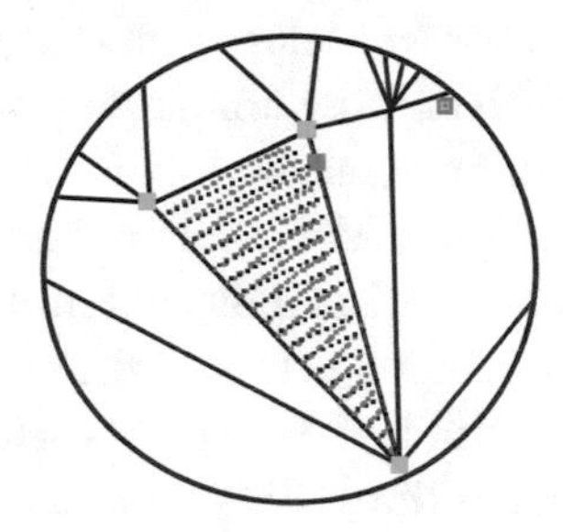

Figure 4.5: Offsets for different landing positions in a triangle, between barycentric coordinates interpolation (black dots) and blending the whole skeleton using SLERP (blue dots).

4.3.5 INVERSE KINEMATICS

An analytical IK solver modifies the leg joints in order to reach the desired position at the right time with a pose as close as possible to the original motion capture data. For footstep-based control, the desired foot position is already encoded in the footstep trajectory, and for COM trajectories the final position is calculated by projecting the current position of the foot over the (possibly uneven) terrain. The controller feeds the IK system with the end position and orientation for each footstep.

4.4 INCORPORATING ROOT MOVEMENT FIDELITY

In some scenarios the user may be more interested in following root velocities than in placing the feet at exact footsteps or with specific walking styles. We present a solution to include root movement-based interpolation into the barycentric coordinate-based interpolator through a user controlled parameter λ_4.

To this end, we incorporate the locomotion system presented by Johansen [105] to produce synthesized motions that follow a COM trajectory with correction for uneven terrain. During offline analysis, a parametric space is defined using all the root velocity vectors extracted from the clips in the motion database. For example, a walk forward clip at 1.5 m/s, and a left step clip at 0.5 m/s produces a parametric space using the root velocity vectors going from the forward direction to the 90° direction, and with speeds from 0.5 m/s to 1.5 m/s.

Given a desired root velocity we define a parametric space Ω_r, and a gradient band interpolator in polar space [105] is created to compute the weights for each animation clip to produce the final blended result. The gradient band interpolator does not ensure accuracy of the produced parameter values but it does ensure smooth interpolation under dynamically and continuously changing parameter values, as with a player-controlled character. Once the different clips are blended with the computed weights, the system predicts the support foot position at the end of the cycle and projects it on the ground to find the exact position where it should land.

The root movement interpolator will select a set of k animations A_1^r to A_k^r with their corresponding weights: $w_1, ..., w_k$. Each of those animations provides a landing position $p_1^r, ..., p_k^r$, and if we only interpolated these animations we would obtain the landing point r.

In order to incorporate the output of the polar gradient band interpolator in the barycentric coordinates-based interpolator we proceed as indicated in Algorithm 1.

The algorithm first checks whether a vertex of the current triangle $\langle p_1, p_2, p_3 \rangle$ can be replaced by any of the three vertices with highest weights selected by the polar band interpolator, p_j^r, $j \in [1, k]$ (lines 1–13 in the algorithm). This replacement happens if the distance between the two landing positions p_i and p_j^r is within a user input threshold ϵ (line 7), and the resulting triangle still contains the desired landing position q_j (function *IsInTriangle* returns true if q_j is inside the new triangle). This means that there is another animation that also provides a valid triangle and has a root velocity that is closer to the input root velocity.

Next, function *CalculateRootLanding* computes the landing position reached after blending the animations given by the root movement interpolator (Eq. 4.3).

$$r = \sum_{i=1}^{k} w_i \cdot p_i^r \tag{4.3}$$

Finally, *ComputeWeights* calculates the three λ_i for the next footstep q_j by incorporating a user provided λ_4 and the result of the polar band interpolator r (Eq. 4.4).

$$q_j = \lambda_1 \cdot p_1 + \lambda_2 \cdot p_2 + \lambda_3 \cdot p_3 + \lambda_4 \cdot r \tag{4.4}$$

and λ_i are defined using the following relationship:

$$\lambda_1 + \lambda_2 + \lambda_3 + \lambda_4 = 1 \tag{4.5}$$

Algorithm 1 Incorporating root movement fidelity

Input:

- The target position q_j,
- The current triangle $\langle p_1, p_2, p_3 \rangle$,
- Root landing positions $\langle p_1^r, ..., p_k^r \rangle$,
- Animation weights $\langle w_1, ..., w_k \rangle \,|w_1 \geq ... \geq w_k$,
- A user input threshold ϵ,
- A user input weight parameter λ_4

Output: $\lambda_1, \lambda_2, \lambda_3$

1: **for** $i = 1$ **to** 3 **do**
2: $\quad u \leftarrow (i+1) \mod 3$
3: $\quad v \leftarrow (i+2) \mod 3$
4: $\quad j \leftarrow 1$
5: $\quad replaced \leftarrow$ false
6: $\quad$ **while** $j \leq 3 \wedge \neg replaced$ **do**
7: $\quad\quad$ **if** $\left\| p_i - p_j^r \right\| \leq \epsilon \wedge IsInTriangle\left(q_j, \left\langle p_j^r, p_u, p_v \right\rangle\right)$ **then**
8: $\quad\quad\quad p_i \leftarrow p_j^r$
9: $\quad\quad\quad replaced \leftarrow$ true
10: $\quad\quad$ **end if**
11: $\quad\quad j \leftarrow j + 1$
12: $\quad$ **end while**
13: **end for**
14: $r \leftarrow CalculateRootLanding\left(\langle p_1^r, ..., p_k^r \rangle, \langle w_1, ..., w_k \rangle\right)$
15: $\langle \lambda_1, \lambda_2, \lambda_3 \rangle \leftarrow ComputeWeights\left(\langle p_1, p_2, p_3 \rangle, \lambda_4, r\right)$

Since w_i and p_i^r are known $\forall i \in \{1, ..., k\}$, and λ_4 is a user input, we have a linear system, where λ_4 determines the trade-off between following footsteps accurately (if $\lambda_4 = 0$), and simply following root movement (if $\lambda_4 = 1$).

As the user increases λ_4 there will be a value $\beta \in [0, 1]$ for which λ_1, λ_2 or λ_3 will be negative, when solving the system of equations formed by Eq. 4.4 and Eq. 4.5. In order to avoid animation artifacts it is necessary to deal only with positive weights, therefore we guarantee that the system will only reproduce q_j accurately as long as $\lambda_4 < \beta$. If we further increase λ_4 beyond the value β then the algorithm will provide the blending values that correspond to a new point q' which is the result of a linear interpolation between q_j and point r. When $\lambda_4 = 1$ the resulting blending will be exclusively the one provided by the root movement trajectory since $\lambda_1 = \lambda_2 = \lambda_3 = 0$. Figure 4.6 illustrates this situation.

Time Warping. Incorporating root velocity in the interpolation does not always guarantee that the time constraints assigned per footstep will be satisfied. Therefore once we have the final set

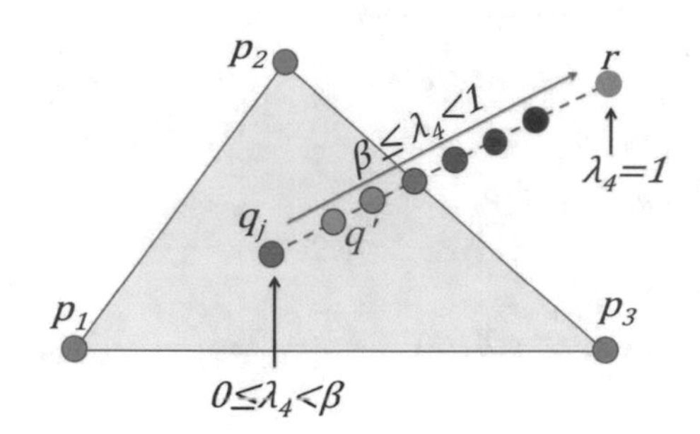

Figure 4.6: When solving the system of equations given by Eq. 4.4 and Eq. 4.5, the value of either λ_1, λ_2, or λ_3 will be negative when $\lambda_4 \geq \beta$. Therefore we need to calculate the barycentric coordinates for a new point q'which moves linearly from q_j to r as the user increases the value of λ_4 from β to 1. This means solving the system of equations for q' instead of q_j, as it is the closest point to the desired landing position which guarantees that all weights in Eq. 4.5 will be positive.

of animations to interpolate between, with their corresponding weights λ_i, $i \in \{1, 2, 3\}$ and w_j, $j \in [1, k]$, we need to apply time warping. Each input footstep f_m has a time stamp τ_m indicating the time at which position q_m should be reached (where $m \in [1, M]$ and M is the number of footsteps in the input trajectory). The total time of the current motion, T can be calculated as the weighted sum of the time of the animation steps being interpolated: $T = \sum_{i=1}^{3}(\lambda_i \cdot t(A_i)) + \sum_{j=1}^{k}(w_j \cdot t(A_j))$. Therefore the time warping factor that needs to be applied can be calculated as: $warp_m = (\tau_m - \tau_{m-1})/T$.

Outside the Convex-Hull. The footstep parametric space defines a convex-hull delimiting the area where our character can land its feet. When a target footstep position falls inside this area, clips can be interpolated to reach that desired position. But if it falls outside this convex-hull we still want the system to consider and try to reach it. Our solution to handle this problem consists of projecting orthogonally the input landing position q over the convex-hull to a new position q_{proj}. Our system then gives the blending weights for q_{proj} and applies IK to adjust the final position. We include a parameter to define a maximum distance for the IK to set an upper limit on the correction of the landing position. By this scheme, even if the input trajectory has some footsteps that are unreachable with the current database of animation clips, our system will provide a synthesized animation that follows the input trajectory as closely as possible until it recovers and catches up with future steps in the input trajectory. This situation is similar to a scenario where the user increases λ_4 and then reduces it again.

4.5 RESULTS

These algorithms have been implemented in C# using the Unity 3D Engine [285]. The footstep trajectories used to animate the characters are generated using the method described in [253] or are created by the user. Some difficult scenarios, exercising careful footstep selection, are shown in Fig. 4.1 and Fig. 4.7. Agents carefully plant their feet over pillars (Fig. 4.7-a) or use stepping stones to avoid falling into the water (Fig. 4.7-b). We animate over a hundred agents at 13 FPS in Fig. 4.7-c and Fig. 4.9.

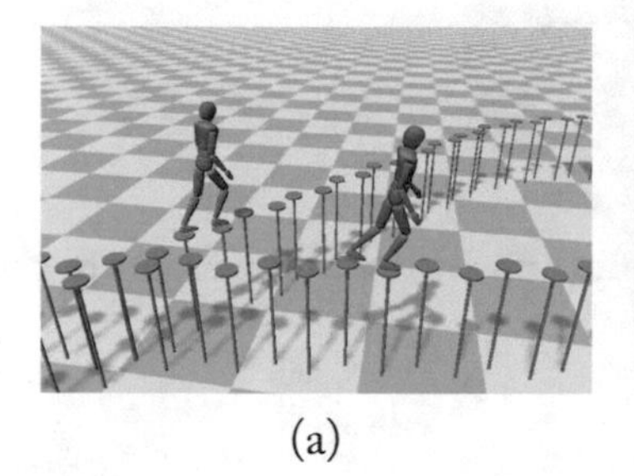
(a)

(b)

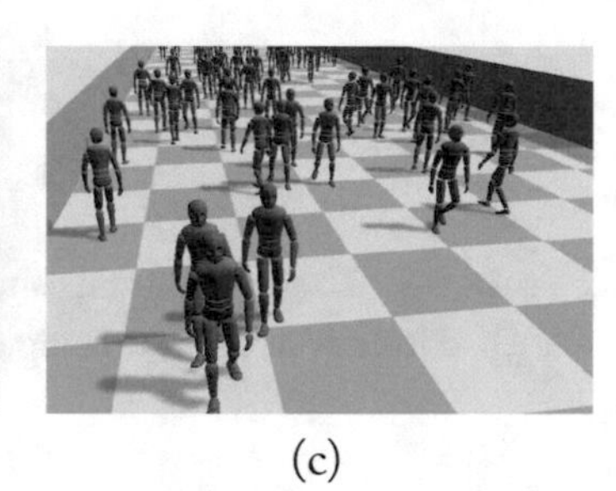
(c)

Figure 4.7: (a) Agents accurately following a footstep trajectory and avoiding falls by carefully stepping over pillars. (b) The stepping stone problem is solved with characters avoiding falls into the water. (c) A crowd of over 100 agents simulated at interactive rates.

Obstacle Course. We exercise the locomotion dexterity of a single animated character in an obstacle course. The character follows a footstep trajectory with different walking gaits, alternating running and walking phases (Fig. 4.1-a,b), and including sidesteps (Fig. 4.1-c) and backward motion (Fig. 4.1-e).

Stepping Stone Problem. Stepping stone problems (Fig. 4.7-b) require careful footstep-level precision: constraints require the character to place its feet exactly on top of the stones in order to successfully navigate the environment. Our framework can be coupled with footstep-based controllers to solve these challenging benchmarks.

Integration with Crowd Simulator. We integrate our animation system with footstep-based simulators [253]; our character follows the simulated trajectories without compromising its motion fidelity while scaling to handle large crowds of characters (Fig. 4.7-c).

4.5.1 FOOT PLACEMENT ACCURACY

The barycentric coordinates interpolator assumes a small offset between the results of linearly interpolating landing positions from the set of animations being blended, and the actual landing position when calculating spherical linear interpolation over the set of quaternions. This small offset depends on the area of the triangle, so as we incorporate more animations into our database, we obtain a denser sampling of landing positions and thus reduce both the area of the triangles and the offset. We believe this is a convenient trade-off since such a small offset can be eliminated with a simple analytical solver but the efficiency of computing barycentric coordinates offers great performance. It is also important to note that if exact foot location is not necessary, and the user only needs to indicate small areas for stepping as in the water scenario, then it is not necessary to apply the IK correction. Figure 4.8 shows the offset between the landing position and the footstep. The magnitude of the error is illustrated as the height of the red cylinders that are located at the exact location where the foot first strikes.

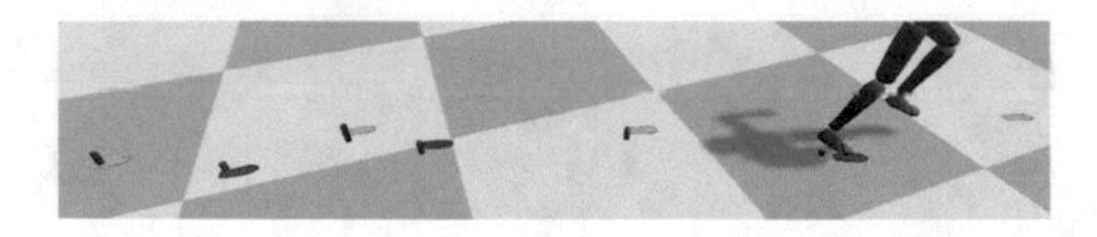

Figure 4.8: The red columns show the small offset between landing position and the footstep when the IK corrections are not being applied.

4.5.2 PERFORMANCE

Figure 4.9 shows the frame rate we obtain as we double the number of agents. Increasing the number of animations would enhance the quality and accuracy of the results, with just a small overhead on the performance.

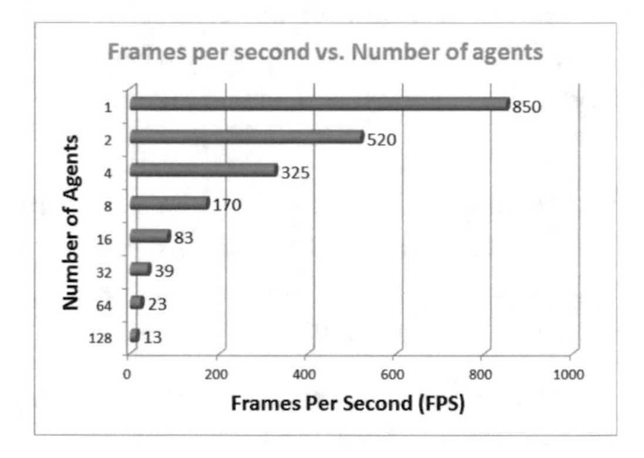

Figure 4.9: Performance of the Footstep Locomotion System in frames per second as the number of agents increases.

The average time of the locomotion controller is 0.43ms, which includes blending animations, IK, the polar band interpolator and the barycentric coordinates-based interpolator. The computational cost of our footstep interpolator is 0.2 ms, which is amortized over several frames as the interpolation in Ω_{f_L} or Ω_{f_R} only need to be performed once per footstep. This time is divided between computing the root movement polar band interpolator which takes 0.155ms and our barycentric coordinates interpolator which takes 0.045ms. Performance results were measured on an Intel Core i7-2600k CPU at 3.40GHz with 16GB RAM.

The bottleneck of the current system is in the root movement polar band interpolator. Firstly because it takes over 77% of the time required to compute the weights, and secondly and most importantly because in many situations, the solution includes a large number of animations with small weights. Since the cost of blending depends strongly on the number of animations being interpolated, it is important to keep this number as low as possible. Simply ignoring those animations with small weights can lead to a final animation that does not fully satisfy the initial constraints. Therefore as future work it would be interesting to consider 3D space for the barycentric interpolator, where the third dimension would correspond to the root velocity. This would allow us to choose the best tetrahedron to satisfy both feet and root constraint, while limiting the interpolation to a maximum of four animations.

CHAPTER 5

Context-sensitive Data-driven Crowd Simulation

Cory D. Boatright, Mubbasir Kapadia, Jennie M.Shapira, and Norman I. Badler

5.1 STEERING IN CONTEXT

Crowd simulations are increasingly called upon for real-time virtual experiences. Allowing users to inhabit such interactions adds additional unpredictability to the virtual agents' decision-making process. Assumptions such as the reciprocity of steering algorithms are not viable in the presence of human input. Thus predicting *a priori* the possible situations an agent will encounter in a virtual world is rapidly becoming intractable as users are given more freedom to control their own avatars. We need algorithms that are scalable not only in agent count, but in circumstance as well.

Data-driven steering algorithms are a natural fit for expanding virtual pedestrians' capability to handle new problems, but current approaches use a single policy as a "one size fits all" approach and are data-bound in their ability to handle general steering. A significant problem for machine learning in crowd steering is the feature space itself, especially for a single-policy system. For behavior as complex and diverse as human steering, increasing numbers of samples can lead to contradictory data which can require an increase in features to try to accommodate the new factor causing the difference. Furthermore, the source of data is often observations of the real world. Data collection poses logistical challenges for gathering adequate, unbiased, and representative samples. These faults ultimately lead to suboptimal scenario coverage in the training data itself. Poor semantic understanding of particular steering choices also inhibits fully robust usage of this data, which can manifest in what appear to be poor steering decisions.

In this section we define the concept of *steering contexts*, which are collections of situations selected for their qualitative similarity. By identifying such contexts, we subdivide the problem space and avoid the single-model problem. We describe a pipeline which leverages these contexts through the use of a collection of machine-learned models trained on synthetic data from a space-time planner. This development pipeline is illustrated in Figure 5.2. Our approach has several advantages over pure data-driven single policy methods. Steering contexts separate data for easier machine learning and allow for scalability of circumstance as well as help mitigate contradictory training samples. Synthetic training data from stochastically generated samples provide less biased data collection resulting in more universal coverage of possible situations. Related publications can be found here [23, 24].

Figure 5.1: Multiple views of a 3,000-agent simulation with high quality rendering.

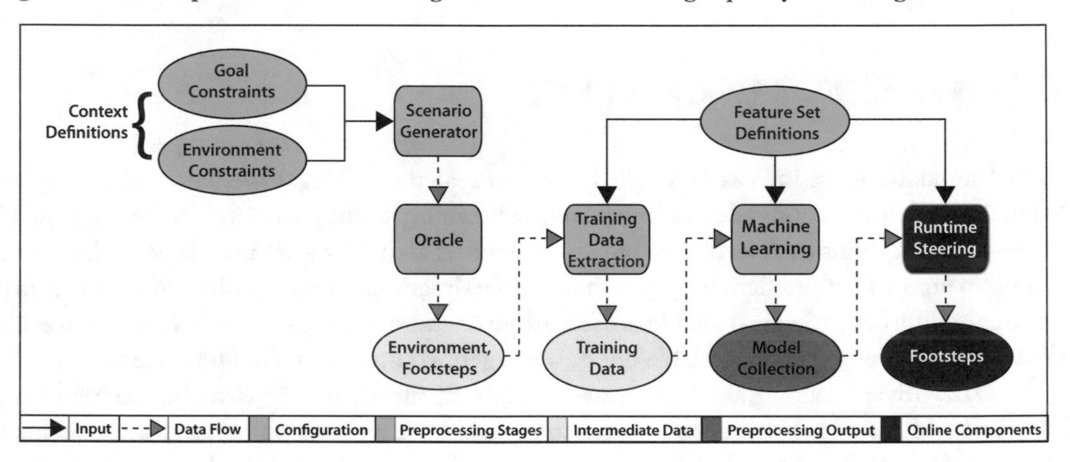

Figure 5.2: Our pipeline for using steering contexts to develop a machine-learned model for use at runtime. The majority of the pipeline is offline processing. A collection of models is trained on data extracted from an oracle algorithm's solution to steering situations, which are stochastically generated. Then each model is a boosted decision tree with its own specialization. The action space consists of footsteps as an advantageous discretization which permits direct control and modeling of human locomotion.

5.2 STEERING CONTEXTS

A **scenario** in scenario space [134] $\mathbb{S}$ is the global configuration of obstacles, agents, and their goals in a virtual environment. The high dimensionality of scenario space makes it inherently intractable to exhaustively cache, so a more general model of steering behavior is needed. Each agent in a frame of simulation encounters its own **situation** S from its perspective. Situations which are similar, based on some feature space $\mathbf{F}^*$, are grouped together to form a **context**, C. The similarity-based grouping is performed to give a high-level perspective on the current situation. To properly steer the agents requires a policy for each context. While these could be manually derived rulesets, identifying contexts and creating policies for each one quickly becomes infeasible. We use machine learning to offset this burden and automatically generate models to serve as a policy for each context.

For a given feature space $\mathbf{F}^*$, context space $\mathbb{C}$ is a projection of $\mathbb{S}$ onto the coordinate system of $\mathbf{F}^*$ and may consist of many overlapping contexts each with boundaries defined by various values of the features. An individual context $C_i \subseteq \mathbb{C}$ is defined in Equation 5.1 with respect to the success of steering policy i in handling situations. A policy is successful if it can produce a valid action from action space $\mathbb{A}$ for the situation, which is one where a collision does not occur and the overall scenario does not deadlock. A scenario then can be considered a sequence of situations and actions with some transition function $\delta(S, a)$.

$$C_i = \left\{S \in \mathbb{C} \mid \exists a : \langle \mathbf{f}, a \rangle, \mathbf{f} \in \mathbf{F}^i, a \in \mathbb{A}, S \neq \delta(S, a)\right\} \tag{5.1}$$

A situation is guaranteed membership in at least one context because, in the worst case, it could have its own special-case policy. This lets us redefine scenario space as $\mathbb{S} = \bigcup_i C_i$, which yields insight into the pursuit of generalized steering. While it would be convenient to know if a set of policies exists that provide optimal behavior for all scenarios in $\mathbb{S}$, this requires the corresponding contexts partition $\mathbb{S}$ based on the "best" context and is thus a direct application of the exact cover problem. Furthermore, it is intractable to know if a set of contexts is sufficient to cover $\mathbb{S}$ as that is an example of the set cover problem. Both of these are known to be NP-Complete [137]. We must therefore approximate contexts rather than strictly define them.

These contexts express how different situations require different policies and improve scenario space by better characterizing regions of success and failure. The approximated contexts identify more constrained domains for data-driven techniques, leading to a more modular and extensible approach to general steering. Examples of contexts we defined by intuition are provided in Figure 5.3 with a full index in Table 5.1. We use a discretization of 4 quadrants of the space surrounding the agent, which we found to be reasonable for the experiments described here. Our work is general and may work for different kinds of context definitions, designed for specific applications.

5.3 INITIAL IMPLEMENTATION

The pipeline for the integration of various contexts into a unified steering algorithm starts with training data, which we generate by means of an oracle algorithm. Next, the various machine-learned models must be fit to the data. Finally, these models are used at runtime to decide where an agent's next footstep should be placed.

5.3.1 TRAINING DATA GENERATION

The feature spaces for learning specialized policies are circular neighborhoods about the agent with discretized wedges that track the nearest agent or obstacle in that region (Fig. 5.4). This geometry

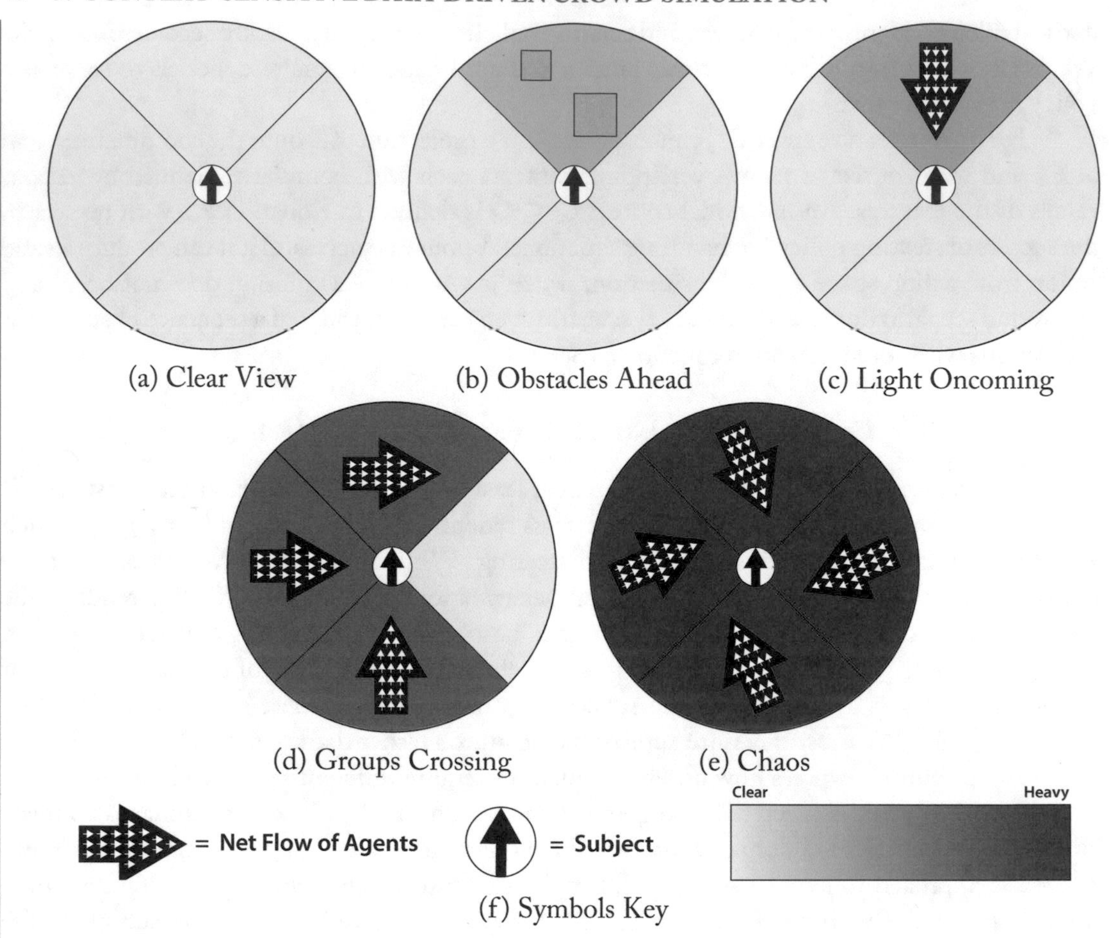

Figure 5.3: Examples from our set of contexts. Net flow is represented by the arrow in each region, density of the region is depicted by darker shades of red, and obstacles are gray boxes. Each of these contexts was stochastically generated with overlap in the permissible values for density. Chaos was generated randomly without regard to any structure as seen in the other contexts. We used a total of 24 contexts.

is partly inspired by the state spaces of [157, 277]. The components of each area are **agent density** and the **net flow** of agents in that area, with the area directly in front of the agent indicating the presence or lack of obstacles. Agent density is a rough approximation of overall crowding in the cardinal directions and includes obstacles. Net flow is the average velocity direction of agents in a particular area. This helps determine whether or not the surrounding crowd is moving with or against the agent, which require different approaches for actions such as collision avoidance. The specialized feature space is a 29-dimensional vector broken down into three values for each

Table 5.1: Parameters which define the 24 contexts we use to prototype our pipeline

Context ID	Obstacles	North		South		East		West	
		Flow	Density	Flow	Density	Flow	Density	Flow	Density
0	Yes	Neutral	Light	Neutral	Light	Neutral	Light	Neutral	Light
1	Yes	Towards	Light	Neutral	Light	Neutral	Light	Neutral	Light
2	Yes	Towards	Medium	Neutral	Light	Neutral	Light	Neutral	Light
3	Yes	Towards	High	Neutral	Light	Neutral	Light	Neutral	Light
4	Yes	Towards	Medium	Towards	Medium	Neutral	Light	Neutral	Light
5	Yes	Towards	Light	Towards	High	Neutral	Light	Neutral	Light
6	Yes	Neutral	Light	Neutral	Light	Towards|Away	Light	Away|Towards	Light
7	Yes	Neutral	Light	Neutral	Light	Towards|Away	Medium	Away|Towards	Medium
8	Yes	Neutral	Light	Neutral	Light	Away|Towards	High	Away|Towards	High
9	Yes	Neutral	Light	Towards	Medium	Away|Towards	Medium	Away|Towards	Medium
10	Yes	Neutral	Light	Towards	High	Away|Towards	Light	Away|Towards	Light
11	Yes	Towards	High	Towards	High	Towards	High	Towards	High
12	No	Neutral	Light	Neutral	Light	Neutral	Light	Neutral	Light
13	No	Towards	Light	Neutral	Light	Neutral	Light	Neutral	Light
14	No	Towards	Medium	Neutral	Light	Neutral	Light	Neutral	Light
15	No	Towards	High	Neutral	Light	Neutral	Light	Neutral	Light
16	No	Towards	Medium	Towards	Medium	Neutral	Light	Neutral	Light
17	No	Towards	Light	Towards	High	Neutral	Light	Neutral	Light
18	No	Neutral	Light	Neutral	Light	Towards|Away	Light	Away|Towards	Light
19	No	Neutral	Light	Neutral	Light	Towards|Away	Medium	Away|Towards	Medium
20	No	Neutral	Light	Neutral	Light	Away|Towards	High	Away|Towards	High
21	No	Neutral	Light	Towards	Medium	Away|Towards	Medium	Away|Towards	Medium
22	No	Neutral	Light	Towards	High	Away|Towards	Light	Away|Towards	Light
23	No	Towards	High	Towards	High	Towards	High	Towards	High

slice: the distance, speed, and orientation of the nearest entity. The distance to the goal and its orientation are the final two values.

A data-driven approach relies on the quality and coverage of its training samples. Real-world data is often used as a source because humans empirically solve any presented steering challenges and we hope to emulate human behaviors. However, we cannot completely control the steering scenarios or know all the variables in the decision-making process of the people observed. To enforce artificial limitations on the scenarios would impact the integrity of the data through the influences of the observer effect. Second, we have no way of knowing *a priori* whether the data set collected has adequate sample coverage for the situations the agents will need to handle. The problem of this potential incompleteness is compounded by the overhead—or impracticality—of collecting additional data. For these reasons, our pipeline uses synthetic data that is readily extensible. All influences are known in advance.

5.3.2 ORACLE ALGORITHM

Our oracle algorithm is based on a memory-bounded A* planner with a discrete footstep action space similar to the action space in [253]. We choose a footstep action space because our machine learning can use classifiers instead of being constrained to regression. When the oracle is run

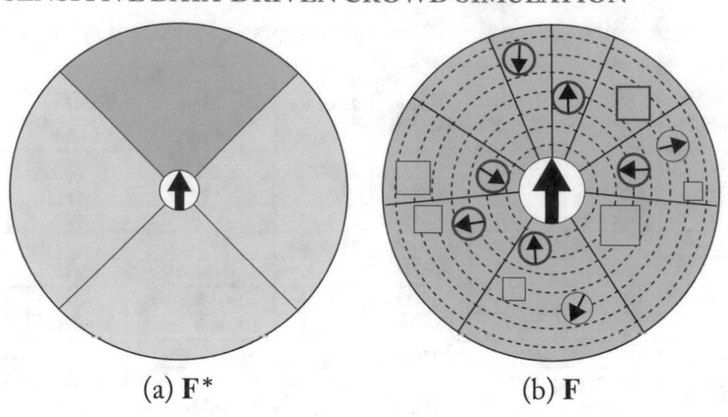

(a) $\mathbf{F}^*$ (b) $\mathbf{F}$

Figure 5.4: The feature sets used in our pipeline, where other agents are circles and static obstacles are depicted as boxes. $\mathbf{F}^*$ is used by the context classifier to dynamically choose the best model based on high-level features, while $\mathbf{F}$ is used to choose the agent's next step based on the local neighborhood.

on the generated scenarios, each agent uses the memory-bounded A* planner to calculate the optimal path from its current location to the goal. The memory bound is raised if a path is not found; as a last resort, Iterative Deepening A* (IDA*) is used. The oracle planner's heuristic used is in Equation 5.2 and is based on the distance to the goal and average expected energy cost to reach that goal.

$$h(\mathbf{p}, \mathbf{g}) = \frac{\|\mathbf{p} - \mathbf{g}\| \cdot \text{energy}_{\text{avg}}}{\text{stride}_{\text{avg}}} \tag{5.2}$$

Each agent has full knowledge only of the obstacles and agents within the horizon of its field of view. Since other agents may enter or leave this field of view, each agent must monitor its path for new collisions and invoke the planner again if such a change is found. The simulations using the oracle are recorded for later extraction of training samples. As the oracle does not use any feature spaces, the same oracle recordings can be used to extract data with different feature spaces, allowing for future exploration of such possibilities. We extract a state-action pair $\langle \mathbf{f}, a \rangle$ where $\mathbf{f}$ is a vector from feature space $\mathbf{F}$ and a are the parameters of the agent's current step, and use it as a sample for training.

5.3.3 DECISION TREES

Avoiding the requirement that the learned policy be a monolithic, universal solution has several key benefits. First, the policies can be simpler and thus executed faster at runtime. Second, we avoid the catastrophically high dimensionality common to such approaches, which are held back by all the factors that can influence every potential action. Finally, we do not need to relearn the entire system to assimilate new data. By using one model to select more specialized models, new

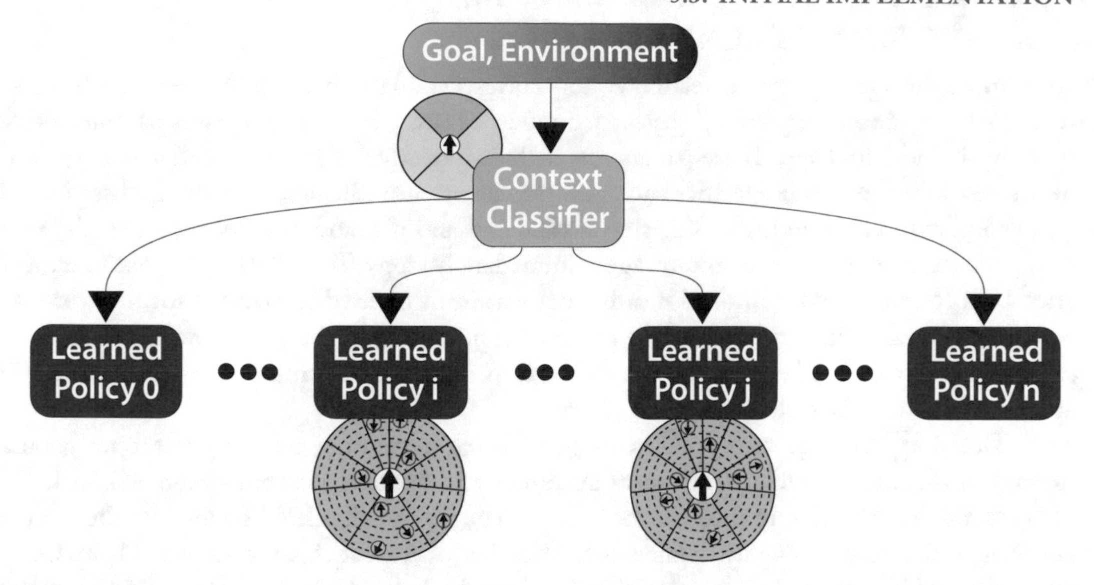

Figure 5.5: The multilevel decision trees used by our models. At runtime the agent gives the model information about its current goal and environment in local-space. This data is used to calculate **f** for each model used. First the context classifier informs the agent of its current context, and the corresponding policy is used to determine the next footstep.

data requires relearning only the particular model it belongs to. Even the creation of a new context only requires the top-level model be recomputed while the other models are still valid and will not be harmed by potentially contradictory data.

This pipeline is agnostic to the specific learning algorithms used at the different levels of the hierarchy, and different algorithms can even coexist on different levels of the hierarchy if that is advantageous to performance. We have chosen to use two levels of boosted decision trees [223] for our instantiation of the pipeline based on the similar problem domain of [273] that showed success for learning different policies that both classified different types of soccer behavior and could be used to decide the actual action itself.

Each of our policies consists of two boosted decision trees; one for each foot. We use a Windows port of the GPL release of the C5.0 decision tree system (`rulequest.com`). We chose ten trees as the amount of boosting empirically based on cross-validation. In total, 2,500 scenarios were sampled from each context and each scenario was generated with respect to a central agent, which provided a variable number of steps per scenario. These steps then became the situations representative of the context for the specialized classifier. A context classification sample was only generated for the first five steps of each recording due to the total number of scenarios that were sampled, all of which supplied data to the context classifier.

5.3.4 STEERING AT RUNTIME

At runtime the agent generates feature vectors corresponding to both the context classifier's feature space and the corresponding specialized model's feature space and receives parameters used to derive its next footstep. These parameters include a relative offset and rotational angle to the next step's location, while specifics such as stride length are calculated on the fly based on the agent's inherent characteristics. This step is validated and if found to be unfit, a default "emergency action" takes place, wherein the agent immediately stops. This allows the agent to try again after a short cool-down period. This safety net was implemented to account for the worst case where a returned action is outside of the parameters permitted by the agents' walking such as two steps in a row from the same foot or too wide a turn. The models cannot be expected to be 100% accurate, which leads to these potential errors.

Decision trees are susceptible to high variance depending on the dataset we generate through stochastic sampling. This causes uncertainty in the decisions our agents will make. We account for this uncertainty through the use of a confidence threshold defined by the C5.0 algorithm. This rating is roughly defined as the number of correct classifications made by the leaf nodes divided by the total number of classifications made by the same node, making it a static quantity once the tree is learned. If the confidence threshold is not met by the classification the agent stops with the ability to resume as conditions change. This confidence value is not a direct reflection on the technique itself, but is instead heavily affected by pruning the decision trees to yield a more general model.

Note that in this algorithm, there is no explicit collision detection or avoidance. In our system, runtime collision detection and avoidance is handled implicitly through the training data itself. This is different from other techniques such as [159] where training samples are used but thorough handling of computed step collisions is required. The steering context training data itself is sufficient to prevent many significant collisions from occurring. Even dense scenarios experience relatively few collisions per capita.

5.4 RESULTS

We generated approximately 2,500 samples for each of our initial 24 contexts. The oracle algorithm required two weeks of continuous computation to return paths for all of the sample scenarios. Those scenarios which were shown to require IDA* were culled in the interest of time. All results were generated on a desktop with 16.0GB of RAM, Intel Core i7 860 CPU at 2.8GHz, and an NVIDIA GeForce GTX 680.

5.4.1 CLASSIFIER ACCURACY

Figure 5.6 plots the error rate for the classifiers used in our experiments. Simulations were run using models trained on amounts of data ranging from 100 to 2,000 scenarios per context. A separate validation set of 200 scenarios per context were kept back to calculate the error rate of the resulting trees.

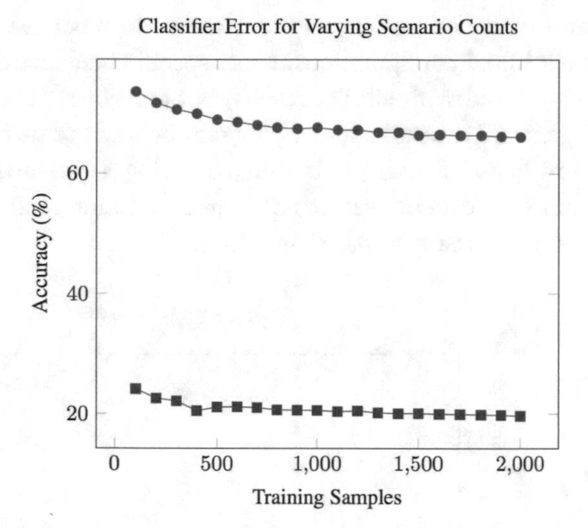

Figure 5.6: Classifier error rates for both context classifier (blue) and an average over the specialized classifiers (red). While the context classifier has a high error rate, a 96% error rate is random chance given the large number of classes to choose from.

Error rates were high but did decrease as data size increased, showing improvement in generalization and not simply noise. Additionally, the average number of steps per context is approximately 12: random guess accuracy would be 8%, which we clearly overcame. Furthermore, random guess accuracy of 24 contexts is 4% which we also surpassed. The error rate seen in the context classifier is likely a result of how the training data was generated in a noisy manner, for instance, some overlap in density between a high density scenario and a medium density scenario exists. A large burden is also placed on the decision trees to distinguish the `Chaos` context from other contexts but this by its nature adds a lot of noise and has no structure, making it difficult to define hyperplanes to separate such scenarios.

5.4.2 RUNTIME

Our instantiation of a context-sensitive pipeline is much faster at runtime than the oracle. As seen in Table 5.2, all contexts experienced speedup, especially significant for the most challenging scenarios involving obstacles. The `Chaos` context, both with and without obstacles, was the most challenging for the oracle and resulted in skewed performance data due to the number of scenarios which were culled. Our method showed an extremely constant amount of time across the different contexts owing to its dynamic model-swapping.

To test the robustness of our collection of models, we created a large-scale simulation consisting of randomly generated obstacles, agents, and goals, as seen in Figure 5.1. This scenario is similar to the `Random` benchmark from SteerBench [251] and serves to challenge the steering algorithm on a wide variety of challenging local interactions between agents and obstacles by randomly sampling their initial configurations. More specific benchmarks designed to test practical use cases were also tested with similar results, but are not reported here for brevity. We measured the time to generate the paths for varying numbers of agents to simulate 1,200 frames, with the results given in Figure 5.7. All tests were run using a single-threaded implementation and real-time framerates were experienced at 1,500 agents and interactive framerates of about 10 FPS were experienced with as many as 3,000 agents.

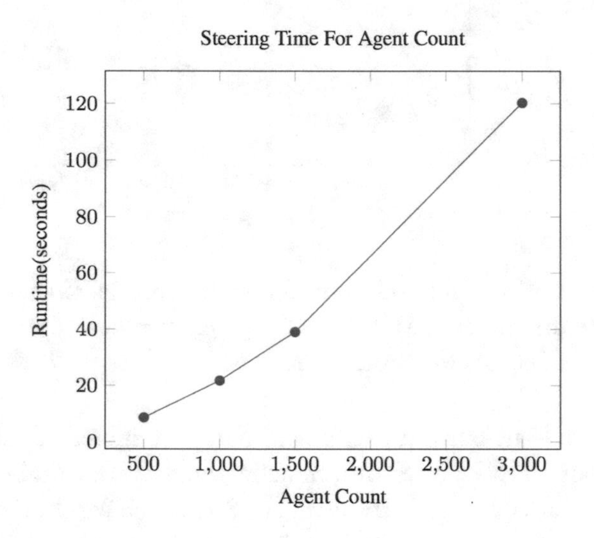

Figure 5.7: Total time taken for computing the steps of a simulation 1,200 frames long for varying numbers of agents with randomly generated obstacles and an overall small area. Overhead was mostly incurred from a naïve implementation of agent density measurement which is $O\left(n^2\right)$ where n is the number of agents.

5.4.3 COLLISIONS

We have run several medium-scale scenarios that are beyond the type of scenarios used for training the models. These scenarios were:

Hallway Two opposing groups of 100 agents cross a hallway.

Random 500 randomly placed agents with 696 randomly placed obstacles throughout the environment.

Table 5.2: Total time for step planning for all contexts in seconds to calculate steps over short scenarios. The first 12 are contexts without obstacles and are based on oncoming and cross traffic patterns with varying levels of agent density. Contexts 12 and above have obstacles and agent patterns matching the upper 12

Context	**0**	**1**	**2**	**3**	**4**	**5**	**6**	**7**	**8**	**9**	**10**	**11**
Oracle	0.73	13.84	5.11	15.53	12.35	9.26	1.68	67.27	101.56	19.90	14.71	1.20
Models	0.07	0.07	0.06	0.06	0.06	0.06	0.07	0.06	0.06	0.06	0.06	0.06
Context	**12**	**13**	**14**	**15**	**16**	**17**	**18**	**19**	**20**	**21**	**22**	**23**
Oracle	123.95	785.0	1945.24	365.25	565.43	574.52	916.30	462.53	3384.10	577.54	396.79	64.78
Models	0.15	0.15	0.17	0.18	0.17	0.18	0.13	0.13	0.15	0.16	0.17	0.16

Urban 2,500 randomly placed agents in an environment simulating an urban area with obstacles as city blocks.

These tests were run for varying numbers of training scenarios, from 100 to 2,000 in increments of 100 and each test was run for 3,600 frames. Afterward, we tabulated the number of collisions and created the graphs in Figure 5.8. The collisions were recorded by severity. Type A collisions have occlusions in the range (0%, 10%] at the worst point. These collisions could be registered due to the circular profile of the agents' bounding volume and thus may not be visible when the simulation is rendered. Type B collisions have occlusion in the range (10%, 35%] and, while more severe than before, could be alleviated with a better anthropomorphic model with torso-rotation. This type of collision is often dealt with in real pedestrians by turning the shoulders to more easily pass one another in cramped conditions. Type C collisions occlude on the range (35%, 75%] and are major collisions which require more tuning to the algorithm to avoid. Type D collisions complete the possibilities at (75%, 100%] and would most likely need a fully reactive collision avoidance system to prevent.

The results were counterintuitive at first. As training samples grew in quantity, so did collisions and even the severity of the collisions. We hypothesize two main factors behind this increase. First, the oracle algorithm is collision-free. Thus a sort of "event horizon" was established in the training data where no reaction to an agent occurs once the agent is too close to another. This means once two agents are too close, there is no impetus to push them apart, which explains the increased number of more serious collisions compared to the more minor offenses.

The second factor is that with increased sample counts, the models better attempt the mimicry of the oracle algorithm's behavior. The oracle has the ability to steer agents together in a very tight, close-call manner. While this is good for the oracle and such nearby passing can be accommodated by it, as the training data increases in size and the agents steer more like the oracle, a mis-step is more likely to cause a collision. In essence, more training data made the agents attempt to steer in a more precise manner, but the inherent inaccuracy of any machine learning algorithm simultaneously leads to higher risk. Thus a collision avoidance process that is invoked on these occasions would be necessary for a data-driven approach to steering.

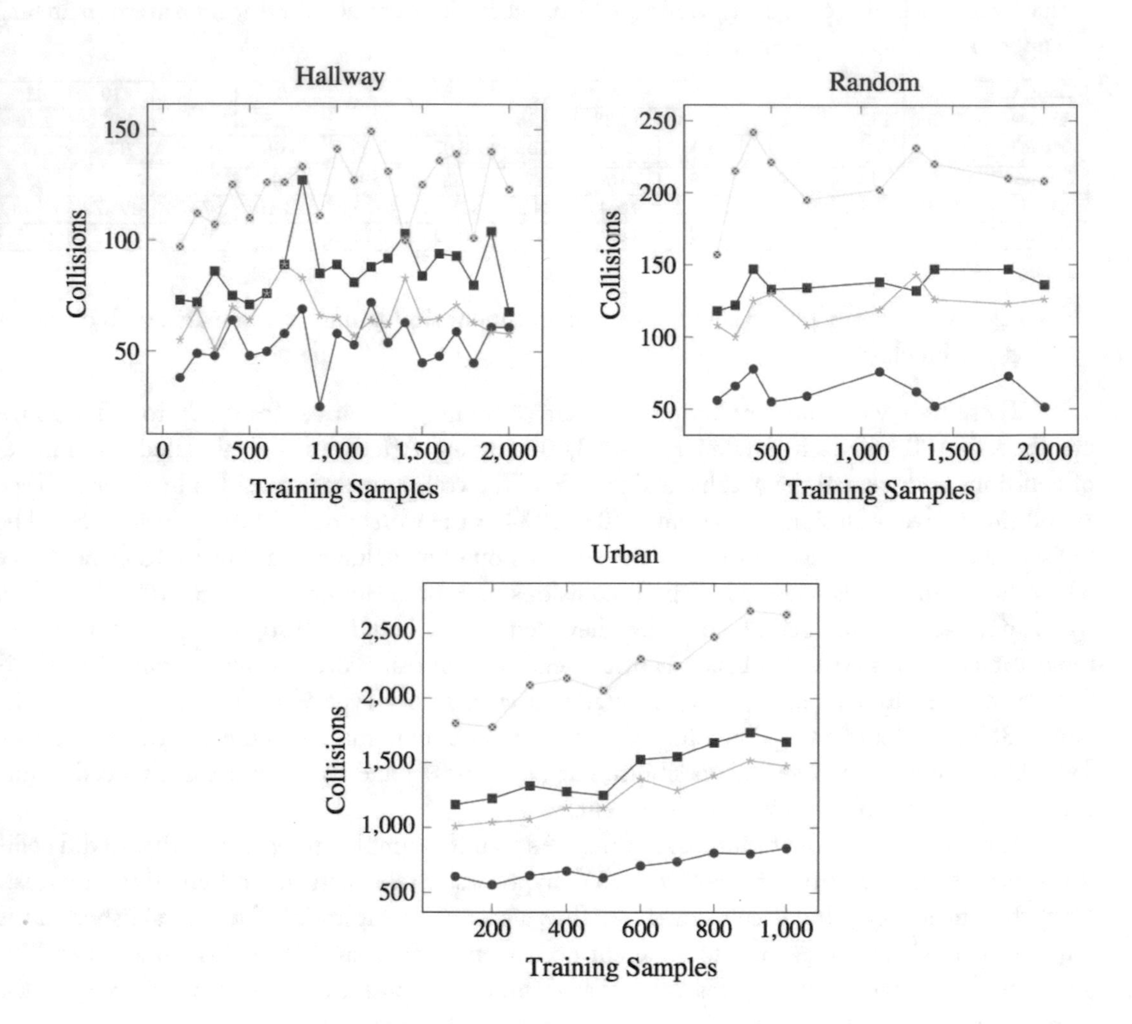

Figure 5.8: Counts for collisions in 3-minute simulations in different test scenarios. Type A collisions are blue, Type B collisions are red, Type C are yellow, and Type D are green. Once collisions occurred, there was little pressure for agents to move apart as the training data was collision-free, thus no samples existed for overlapping agents. Note that while high, per capita an agent in each of these simulations is only likely to encounter around 1–3 collisions with approximately one third of them minor in nature in spite of the lack of any explicit collision avoidance.

CHAPTER 6

Conclusion

6.1 FOOTSTEP-BASED COLLISION AVOIDANCE

Footsteps are a natural locus of control since they are the major contact points between a bipedal system and the external environment. However, there are some prominent aspects of bipedal locomotion that are important in specialized situations. For example, specialized knee and ankle bending motions are fundamental for starting and stopping movements as well as going up inclines and stairs. Furthermore, the center of pressure (i.e., pendulum pivot) shifts from heel-to-toe during a step, and humans can rotate their foot orientation in place, improving locomotive efficiency. Angular momentum is another important physical quantity that has importance for dancing and athletic activities. All these aspects may be important for a "universal" virtual character, and so it is appropriate to consider how to bring these aspects into the simulation.

Not all costs that a character evaluates are related to effort or energy: there are likely significant costs related to social and cognitive behaviors, rather than minimizing effort. We found it curious that foot orientation did not have to be a parameter in order for the planner to find natural walking footsteps. This leads to an interesting biomechanics question: what are the actual abstractions and parameters real humans use to guide footstep selection? Our work offers one candidate answer, but more rigorous studies can be done in this direction. Our work considers each agent as an individual who plans its own footstep trajectories for collision-avoidance. For more challenging scenarios, we must consider centralized approaches where interacting agents may coordinate to resolve potential deadlocks. Chapter 9 describes a multi-domain planning approach that models different resolutions of decision-making for planning agents in crowded situations.

6.2 FOOTSTEP-BASED LOCOMOTION

We have described a system that uses multiple parameter spaces to animate fully embodied virtual humans who accurately follow a footstep trajectory respecting root velocities, using a relatively small number (24 in our examples) of animation clips. This method is fast enough to be used with tens of characters in real time (25 FPS) and over a hundred characters at 13 FPS. The method can handle uneven terrain, and can be easily extended to additional locomotion behaviors by grouping new sets of animation clips and generating different parametric spaces. For example, walking and running motions can be blended together, but if we wanted to add crawling or jumping motions, it would be better to separate them into different parametric spaces for each style. This will avoid unnatural interpolations that can appear when blending between very different styles. Having different parametric spaces requires some sort of classification based on motion characteristics, such as changes in acceleration, maximum height of the root, length of flight phase, etc. Assuming we can extract the parametric spaces for different animation types, we would also need additional transition clips to switch between disparate locomotion types such as crawling and walking.

We do not run physical or biomechanical simulations, instead relying on interpolation and blending between motion capture animations. Accuracy thus depends on the variety of animation clips, while quality and efficiency depends on the number of clips. Our method strikes a good equilibrium in the trade-off between efficiency and accuracy.

In order to reduce the dimensionality of the footstep problem, the orientation of the previous footstep is not included in our parametric space. Ignoring the final orientation of the character at the end of the previous step can induce some discontinuities between footsteps. We mitigate this effect by blending between footsteps automatically for a small amount of time (about 0.2 seconds) to reduce the computation cost and improve scalability for larger groups of agents.

For future work we could extend the barycentric coordinate interpolator to 3D space with the third coordinate being the root velocity. This will free our system from the polar band interpolator which not only takes longer to compute but also selects too many animations, resulting in slower blending. Another option could be to interleave the execution of the footstep-based locomotion controller for different characters in different frames, ensuring we do not execute it for all the agents in each timestep.

6.3 CONTEXT-BASED STEERING

We have defined steering contexts, a new view on the space of possible scenarios an agent may encounter as it steers through its virtual world. These contexts provide new insight into the task of creating a general steering controller capable of handling any situation it encounters. Unless the controller can be independently proven to be general and thus consist of a single context, the algorithm will fragment scenario space into subsets which must each be handled by a separate policy. This creates a coverage uncertainty that is by nature NP-Complete and to our knowledge no real-time algorithm is unaffected by this discovery.

We have also proposed a pipeline for constructing a steering algorithm that is both context-sensitive and scalable to circumstance. Through the use of a multiplicity of models fit to steering contexts, machine learned policies can be combined for better and more structured coverage of the space of possible scenarios than would otherwise be commanded by a single-model approach generalized to all situations. We used an oracle algorithm to get high quality, on-demand training data which can be used for new contexts without the overhead, bias, or uncertainty of real-world data. This training data was then broken into contexts based on intuition and policies fit for each context using machine learning.

We noticed that the decision tree models used to prototype our pipeline are too restrictive if the chosen action is incorrect. A naïve Bayesian approach would allow a better "next best" progression of footstep selection rather than the current all-or-nothing approach. Multiple algorithms can coexist throughout the collection of policies allowing each context to be fit as needed for better overall accuracy. Furthermore the contexts themselves could be defined from a collection of data using unsupervised clustering, further removing the human element from the problem.

Currently we decide the next step an agent should take and deciding multiple steps would require an exponential increase in the size of the action space if done naïvely. However, we postulate that analysis of step sequences would reveal that not all step combinations need to be learned, drastically decreasing the overhead. Maneuvers such as overtaking other pedestrians or rounding corners could then be encapsulated, rather than depending on each step in the process to be determined individually. Even with 90% decision accuracy, a 5-step sequence has a probability of being correct of only about 60%. An alternative to the short horizon of a single step could be to influence navigation instead and compute a waypoint, which a fast but reactive algorithm uses to guide the agent's path. This observation provides a natural segue to the path navigation problem which we tackle next. Finally, the methods described in this paper are agent-centric approaches for goal-directed collision avoidance. Modeling the dynamics of social groups and multi-actor interactions requires additional layers [100, 239] that orchestrate cooperative behaviors between multiple agents.

PART II

Multi-agent Navigation

CHAPTER 7
Background

Autonomous agents navigating virtual environments need a robust abstract representation of the walkable or "free" space. Deciding the most adequate representation—the navigation mesh (NavMesh) [258]—can be a challenge with strong dependencies on the application and the size of the crowd. Ideally the navigation mesh should be formed by cells with a tight fit to the environment geometry. Navigation meshes based on 2D regular grids do not generally satisfy such a condition, though they can provide great computational advantages when calculating paths for multiple agents in real time.

In [208], we presented a framework to semi-automatically generate cell and portal graphs (CPG) for environments described by geometry with multiple floors connected by stairs, but limited to axis-aligned walls and discrete door widths. The user could readily define an environment through a text editor and a library of symbols, and the system would automatically generate both the 3D geometric environment and its cell and portal navigation mesh.

Automatically determining navigation meshes in arbitrary geometric spaces, however, is a complex problem, and it is often done by hand or interactively. In Chapter 8 we present a novel framework to compute navigation meshes fully automatically for large and complex 3D environments.

Rather than develop a single "best" space representation, Chapter 9 describes a multi-domain framework that exploits a variety of representations to speed up crowd simulation. Having a multi-domain system can also leverage different kinds of information during path planning. For example, planning highly detailed trajectories taking into account all the features of the environment (e.g., static obstacles, the dynamics of moving objects, other agent's decisions, etc.) can be extremely costly. Computing paths for many agents in real time needs to consider as few variables as possible. The key is to find a good equilibrium that can provide the best of both strategies.

7.1 NAVIGATION MESHES

The determination and representation of free space in a virtual environment is a central problem in the fields of robotics, videogames, and crowd simulation [112]. Two general approaches to representing free space are roadmaps and cell decompositions. The main objective of both methods is to generate a graph that can be used by a search algorithm (usually A*) to find a path free of obstacles between two points in the scene.

The roadmap approach [5, 230, 264, 305] captures the connectivity of the free space by using a network of standardized paths (lines, curves). The main limitation of this representation is that it only contains information about which locations of the scene are directly connected, but it does not describe the geometry nor obstacles in the scene. Consequently, avoiding dynamic obstacles is usually difficult and not always possible [264].

The cell decomposition approach partitions the navigable geometry of the scene into convex regions, guaranteeing that a character can move between two points in the same cell following a straight line, without getting stuck in local minima. This particular decomposition is usually called a navigation mesh, from which a CPG can be obtained to compute paths free of obstacles [258]. Collisions with movable obstacles such as other agents are solved by using a local movement algorithm [207, 226], or by dynamically modifying the NavMesh [113].

Local movement algorithms are generally driven by setting waypoints within the portals of the NavMesh that work as attractors to steer the agents in the right direction. An improvement to traditional waypoints was introduced by [41] who used the whole length of the portal to attract the local movement of the agents, thus resulting in more natural looking paths.

Uniform grid cells can be used to partition the environment [10, 146]. It is an easy and fast solution to obtain a convex decomposition, but the main problem of this technique is that a high-density CPG is generated, increasing the time of the search algorithm. Another limitation is the poor adjustment to the shape of the obstacles, thus losing important parts of the walkable space. Valve's Game Engine follows a similar approach [286]. It subdivides the virtual map by axis aligned (AA) quads of arbitrary size. The resulting partition contains fewer cells than in the case of uniform grids, but is still far from optimal as it is a partition restricted by AA quads. Hale et. al. [83] presented an automatic NavMesh generator method that consists of spreading a certain number of unitary quad seeds in the scene. Those quads are expanded as much as possible, adjusting to the contour of the obstacles and generating new seeds to completely cover the walkable space. Although the adjustment to the obstacles is perfect, the partition obtained contains many narrow cells and T-Joints, which could introduce artifacts in the movement of the characters during steering. The method only works if every obstacle is convex, so obstacles must be preprocessed into convex parts. A volumetric version of this algorithm in [82] has the same limitations as the 2D version. Recent approaches have successfully harnessed the power of GPUs to speed up search efforts on uniform [122] and adaptive resolution grids [69].

Triangular meshes have also been used to represent a NavMesh. In [111, 113], a dynamic constrained Delaunay triangulation (CDT), having as constraints the edges of the obstacles, is used to represent the walkable area of a scene. At run-time, obstacles may be inserted, removed, or displaced and the CDT is able to dynamically accommodate these changes. In practice, however, application performance greatly depends on the complexity of the CDT, as well as on the complexity and number of constraints being moved. The size of the CDT is linear in the input vertices, while the convex partition obtained by our approach (described below) is linear in the number of concave vertices and thus our method scales better. In [112], Kallmann introduced a new representation called a local clearance triangulation (LCT) that allows computing paths free of obstacles with arbitrary clearance. Such a triangulation is obtained by a process that iteratively refines the starting CDT of the scene. These methods have also been applied for planning paths for coherent and persistent groups [100].

TopoPlan [153] is an application that automatically generates a CPG for a given virtual environment defined as a mesh of triangles that can contain multiple layers. The layers are extracted using a prism subdivision, and the CDT of each layer is computed. Although the description of the walkable space is perfect, it is very costly to compute. Results show that for a scenario of just 120k triangles, Topoplan needs more than 15 minutes to compute the NavMesh. This work was then extended to identify outdoor, indoor, and covered areas for spatial reasoning [107]. By using a partition based on triangles, geometric operations are very efficient, the convexity of the partition is guaranteed, and it contains the least possible number of ill-conditioned cells. However, the partition is far from optimal as it is restricted to triangles.

Another representation consists of partitioning the space by using Ngons [189, 199, 278, 284]. An example of such an application is the Recast toolkit [189], a popular tool used in videogames and other simulations. The method used by Recast is inspired by the work of Haumont et al. [86]. First it computes a voxelization of the scene that makes the method robust against degeneracies of the input model (such as interpenetrating geometry, cracks or holes) and simplifies any furniture. Then a partition of the scene is obtained by applying the watershed transform (WST) on the distance map field of the voxelized scene. This partition is further refined as the WST does not guarantee convexity of the generated cells. Recast is a robust application, but the description of the walkable space is not totally exact as the adjustment to the contours is not precise. Additionally, its most important drawbacks are over-segmentation of the space (which increases search cost), and generation of walkable but inaccessible regions (which might need to be removed by manual post-processing).

Toll et al. [295] present a NavMesh generation method for a multi-layered environment, such as an airport or a multi-story car-park, where the different layers of the scene are connected by elements such as stairs or ramps. However, they do not provide an automatic method to extract such layers. The implementation of this NavMesh generation method, restricted to one single layer, can be found in [70].

Oliva and Pelechano [199] describe an efficient technique to calculate a convex decomposition with a number of cells close to the optimum. It only works for simple polygons with holes, however, so for a real scene the polygons have to be introduced manually, and the CPGs merged together by hand. In Chapter 8 we describe a practical framework to automatically create NavMeshes from any complex 3D multi-layer environment.

7.2 PLANNING

Given that there is a (possibly dynamic) NavMesh for the environment, the next problem to solve is the determination of appropriate navigation paths for every moving agent. This necessitates search and planning strategies. There is extensive research in multi-agent simulations with many proposed techniques that differ in domain complexity and control fidelity. Global navigation approaches [112, 265, 268, 291] precompute a roadmap of the global environment which is used for making efficient navigation queries, but generally regard the environment as static.

Planning-based control of autonomous agents has demonstrated control of single agents with large action spaces [36, 66, 241]. In an effort to scale to a large number of agents, meet real-time constraints, and handle dynamic environments, a variety of methods [216] have been proposed. The complexity of the domain is made simpler [155, 170] to reduce the branching factor of the search, or the horizon of the search is limited to a fixed depth [35, 253]. Anytime planners [168, 290] trade-off optimality to satisfy strict time constraints, and have been successfully demonstrated for motion planning for a single character [234]. The work in [124, 195] extends anytime dynamic planning to handle spatial constraints. Randomized planners [99, 241] expand nodes in the search graph using sampling methods, greatly reducing search efforts to make it a feasible solution in high-dimensional, continuous domains. The work in [95] exploits the use of graphics hardware to enable interactive motion planning in dynamic environments.

Hierarchical planners [26, 28, 97] reduce the problem complexity by precomputing abstractions in the state space, which can be used to speed up plan efforts. Given a discrete environment representation, neighboring states are first clustered together to precompute abstractions for high-level graphs. Different algorithms are proposed [144] which plan paths hierarchically beginning at the top level, then recursively planning more detailed paths at lower levels, using different methods [150, 263] to communicate information across the levels. These include using the plans in high-level graphs to compute heuristics for accelerating searches in low-level graphs [96], using the waypoints as intermediate goals, or using the high-level path to define a tunnel [75] to focus the search in the low-level graph.

The work presented in Chapter 9 builds on top of excellent recent contributions [160, 171] showcasing the use of space-time planning for global navigation of a single agent in a dynamic environment. Levine et al. [160] uses parameterized locomotion controllers to efficiently reduce the branching factor of the search and assumes that object motions have known trajectories, thus mitigating the need for replanning. Lopez et al. [171] introduces a dynamic environment representation which is computed by deducing the evolution of the environment topology over time, thus enabling space-time collision avoidance with no prior knowledge of how the world changes. In contrast, we use multiple heterogeneous domains of control, and present a planning-based control scheme that reuses plan efforts across domains to demonstrate real-time, multi-character navigation in constantly changing dynamic environments. Instead of automatically computing abstractions from a given representation, we develop a set of heterogeneous domains with different state and action representations that provide trade-offs in control fidelity and computational performance, and investigate different methods of communicating between domains to meet our application needs.

CHAPTER 8

Navigation Meshes

Ramon Oliva and Nuria Pelechano

Character navigation in complex scenes is commonly performed by having a Navigation Mesh (NavMesh), which encodes a convex decomposition of the scene. The NavMesh is represented with a Cell-and-Portal Graph (CPG), where cells are convex regions and portals are the edges shared by convex regions that a character can traverse. Path finding algorithms such as A* are then used over those NavMeshes.

Although NavMeshes are widely used in video games and virtual simulations, there are limited applications that automatically generate a NavMesh suitable for path planning. Either the user refines semi-automatically generated NavMeshes, or creates them manually from scratch, both of which are time consuming, tedious, and error-prone. Automatic methods to generate good quality CPGs with minimal numbers of cells would be highly desirable.

An effective CPG generator should avoid creating ill-conditioned cells that can lead to artifacts in the characters' local movement. An ill-conditioned cell has vertices that are practically collinear or whose Area/Perimeter ratio is close to zero.

Over-segmented partitions cause problems, too. The performance of the path finding algorithm directly depends on the dimensions of the generated graph: so fewer cells mean faster searches. Depending on the underlying local movement algorithm being used, an over-segmented partition may end up with characters walking in zig-zags through a long convex space as they are forced to go through unnecessary portals. Portals that are too close together may add unnecessary nearby attractors, and therefore complexity, when trying to achieve natural-looking local movement.

8.1 NAVMESHES FROM 3D GEOMETRY: NEOGEN

Our CPG generator, called NEOGEN, inputs a polygon soup describing a multi-layer 3D environment and outputs a CPG that can be directly used for character navigation without any further manual effort. NEOGEN requires four input parameters:

- maximum step height, h_s, indicates the maximum difference in terrain height that the character can overcome.

- maximum walkable slope angle, α_{max}, indicates the maximum angle of the slope that the character can walk up or down.

- height of the character, h_c.
- walkable seed, s_w, specified by the user to indicate one point of the geometry where the characters can navigate.

Note that the first two are character navigation skills and could be set by the user or calculated automatically from the set of animation clips. Figure 8.1 shows some intermediate results provided by NEOGEN, along with the final CPG. Figure 8.2 shows the NEOGEN navigation mesh from a 3D multi-layer environment.

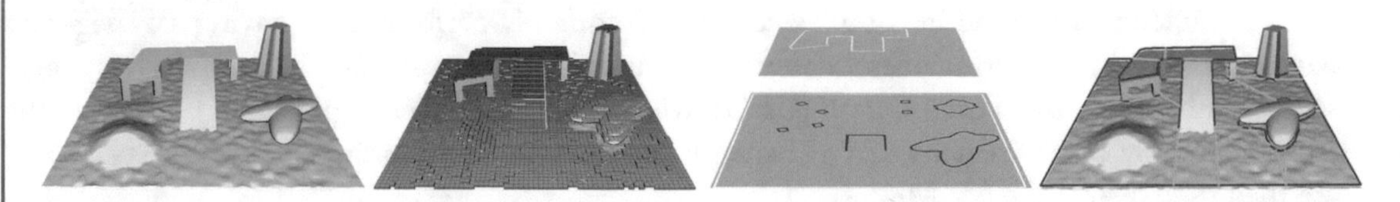

Figure 8.1: Near-optimal navigation mesh construction for a simple example of a 3D environment with 2 layers. From left to right we can see the original scene, the result of the layer extraction step after the coarse voxelization, the 2D floor plan of each layer, and finally the near-optimal navigation mesh.

The first step of the method consists of performing a GPU *coarse voxelization*. Voxels are classified as empty, positive, or negative. Geometry does not need to be axis aligned, and the discretization at this level will not affect the final result. This voxelization is then processed by the *Layer Extraction and Labeling* step to classify the voxels into layers.

Once the layers have been detected, the *Layer Refinement* step performs a high resolution orthogonal render of each individual layer in order to obtain a 2D image of the layer that preserves the original geometry. From this 2D image the floor plan of each layer is computed. Finally a near-optimal convex decomposition of each floor plan is carried out and linked to the adjacent layers in order to obtain the CPG of the scene.

In our method, the voxel size is not used directly to define walkable areas, but to serve as a filter to process the underlying geometry. Therefore the discretization at this level does not have an impact on the adjustment of the cells to the geometry. The only thing that matters is not to have an overhanging walkable area over another within the same voxel, which is guaranteed by having the voxel height equal to the characters' height.

8.1.1 GPU COARSE VOXELIZATION

The voxelization method is an extension of the GPU appraoch in [55], where a voxelization is performed in just one rendering pass based on a slicing method. A grid is defined by placing a camera above the scene and adjusting its view frustum to enclose the area to be voxelized. This camera has an associated viewport with (w, h) dimensions. The scene is then rendered, constructing the voxelization in the frame buffer. A pixel (i, j) represents a column in the grid and each

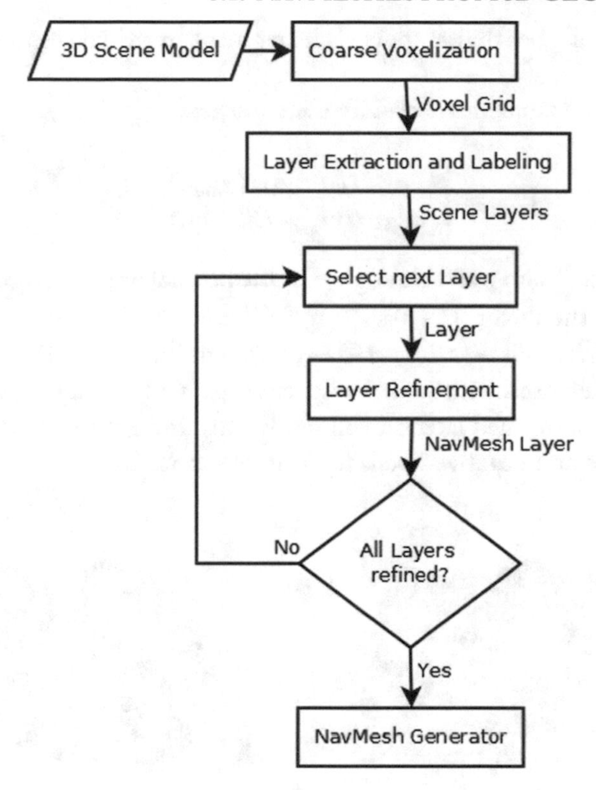

Figure 8.2: Multilayer framework for the automatic navigation mesh generator.

voxel within this column is binary encoded using the kth bit of the RGBA value of the pixel. Therefore, the corresponding image represents a $w \times h \times 32$ grid with one bit of information per voxel. This bit indicates whether a primitive passes through a cell or not. The union of voxels corresponding to the kth bit for all pixels defines a slice. Consequently, the image/texture encoding the grid is called a *slicemap*. When a primitive is rasterized, a set of fragments are obtained. A fragment shader is used in order to determine the position of the fragment in the column based on its depth. The result is then $OR - ed$ with the current value of the frame buffer.

Since faces nearly parallel to the viewing direction do not produce any fragment, we have extended the previous technique by carrying out three separate *slicemaps*, one for each viewing direction (along the X, Y, and Z axis respectively). A final *slicemap* is then created by merging these separate *slicemaps* into a single one.

Each voxel will be of size $w \times w \times h_s$ where w can be a user input, or else can be automatically initialized to the character's radius, and h_s is the maximum step height. Since voxels are classified into positive, negative, or empty, we use two *slicemaps*, one to store positive fragments and the second to store negative fragments. This can be done in a single pass using *Multiple Render*

Target (MRT) and a shader that outputs each fragment in the corresponding texture depending on its classification.

For each drawn fragment we classify the voxel as:

$$V_{ijk} = \begin{cases} positive & \cos(\alpha_{max}) > (\vec{n} \cdot \vec{UP}) \\ negative & otherwise \end{cases}$$

where α_{max} is the maximum walkable angle, $\vec{n}$ is the normal of the polygon corresponding to that fragment and $\vec{UP}$ is the vector $(0, 1, 0)$.

Note that *positive* and *negative* surfaces can fall in the same voxel. Since we are only interested in detecting the voxels with walkable surfaces, we will classify those voxels also as *positive*. The refinement step performed later on will resolve any ambiguity. Figure 8.3 shows the decomposition into positive and negative voxels for a simple scene.

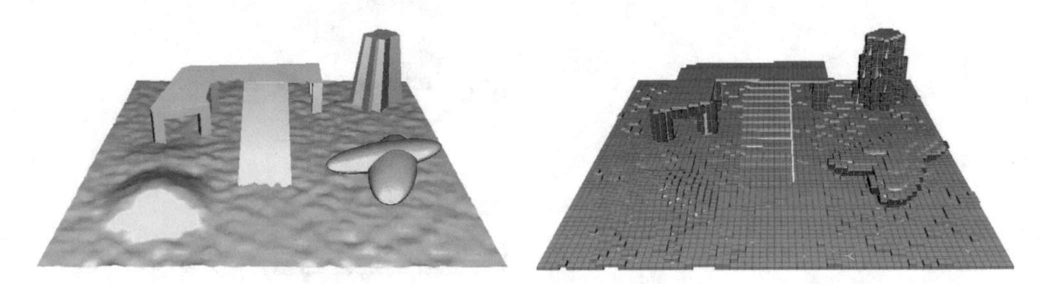

Figure 8.3: Original scene and its corresponding GPU coarse voxelization. Red indicates negative voxels and blue positive voxels (i.e., where the slope is within the walkable capabilities of the character).

Currently the maximum resolution for the coarse voxelization is $128 \times 128 \times 128$. This resolution is limited by having *MRT* that can render into 8 textures in a single pass and we need 2 for each *slicemap* (*positive* and *negative*). Therefore we can count on 4 textures to encode each *slicemap*, with 8 bits/channel and 4 channels (RGB).

The size of the environment that we can currently handle is restricted by the resolution of the voxelization step. However it could easily be extended to larger environments by subdividing the scene into smaller regions to fit with the resolution of the voxelization, treat each of these regions as an individual multilayered map applying the current method, and finally join the resulting NavMeshes into a single one.

8.1.2 LAYER EXTRACTION AND LABELING

After the coarse voxelization step, we know which voxels of the scene contain potentially walkable geometry. The potentially walkable area is formed by all those voxels:

$$W_A = \cup\{V_{ijk} = positive\}$$

In this step, we need to split the potentially walkable area W_A into layers, as well as eliminate all those voxels that are unreachable due to the character's maximum step height h_s, or the character's height, h_c. A layer, L_i, will be composed by a set of *connected accessible* voxels such that it is not possible to have in the same layer V_{ijk} and $V_{ijk'}$ (i.e., it can contain at most one voxel per column of the voxelization). An *accessible* voxel is a *positive* voxel where the character can stand without colliding with any overhead geometry:

$$V_{ijk} = \begin{cases} accessible & V_{ij(k+\{1..n\})} = empty, n = \left\lceil \frac{h_c}{h_s} \right\rceil \\ non\ accessible & otherwise \end{cases}$$

Two *accessible* voxels V_{ijk} and $V_{ijk'}$, (where $k \leq k'$) are *connected* when the distance in both i and j is at most 1 and $V_{ijk''} = positive, \forall k'' = [k+1, k'-1]$, when $k' - k > 1$.

An ordered flooding algorithm is performed to extract all the different layers and assign them layer IDs, L_{id}. Initially *accessible* voxels are stored ordered from bottom to top and assigned an invalid L_{id}. Starting from the most bottom *accessible* voxel, an L_{id} is created and its *connected accessible* voxels are checked for an L_{id} propagation step; there are three possible cases:

- The voxel has an invalid L_{id}: the current L_{id} will be assigned, as long as there is no voxel below it that already has the current L_{id}. Otherwise the voxel will remain with an invalid L_{id} until the flooding method reaches it.
- The voxel has a valid L_{id} different from the current one: Layer merging can be carried out between the two layers if there are no voxels from one layer in the same column as a voxel from the other layer.
- The voxel already has the same L_{id}: in this case nothing needs to be done.

The flooding algorithm proceeds iteratively from bottom to top. Figure 8.4 (left) shows the result of the layer extraction step.

Finally, layers that are unreachable for the seed s_w provided by the user are eliminated. Figure 8.4 (right) shows the result of the layer connectivity step.

8.1.3 LAYER REFINEMENT

At this stage, we have subdivided the real walkable space into layers. The Navigation Mesh could be computed from this representation, but since we want to obtain a close conformance to the obstacles, we need to further increase the resolution. The goal is to obtain a 2D high resolution floor plan of each layer. To do this we use a fragment shader that for each layer will only render the geometry that corresponds to such layer. Once the fine floor plan is rendered we can calculate the polygon bordering the layer, as well as the obstacles (Fig. 8.5).

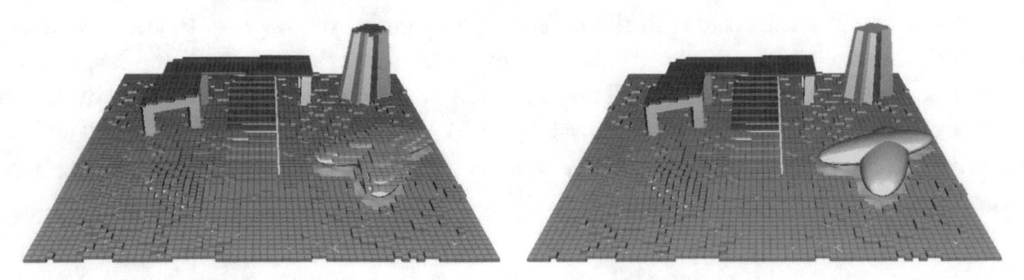

Figure 8.4: On the left, the result of the layer extraction step with each color indicates the set of voxels belonging to the same layer. On the right, the result of the layer connectivity step where unreachable layers (based on character's step height, h_s) have been eliminated.

Layer contour expansion

For each layer obtained from the coarse voxelization we have two types of voxels: *accessible* and *obstacle*. It is possible though that *obstacle* voxel neighbors of *accessible* voxels partly contain walkable geometry. Contour expansion is thus computed in order to consider these voxels for the cutting shape calculation. Figure 8.6 shows the result of this step near obstacles or floating geometry.

Cutting Shape

The *Cutting Shape* (CS) is calculated from the accessible voxels of each layer. The CS wraps each layer in order to filter the geometry that should be rendered into a 2D high resolution texture to obtain the floor plan of a layer. This CS stores for each pixel the depth of the top of the accessible voxels with the offset h_c and the type of voxel. The output texture stores:

- Channel R: the type of voxel (1 for accessible voxel, 0.5 for voxels which contain a portal between layers, and 0 for non-accessible voxel).
- Channels GBA: the depths of the cutting plane for each column of the voxelized grid.

Depth map extraction

An orthogonal top view camera is defined enclosing the scene. The depth map of the layer is calculated with a fragment shader, using the CS as a filter. The fragment shader will discard a fragment (i, j) if it satisfies any of the following conditions:

1. $texture_{ij}^{CS}.R = 0$. If channel R of the cutting shape contains a 0, it means it is a non-accessible voxel.
2. $fragment_{ij}.depth < texture_{ij}^{CS}.GBA$. If the current fragment's depth is smaller than the CS depth (stored in channels GBA) then those fragments do not belong to the geometry of the current layer, but to some other higher layer.

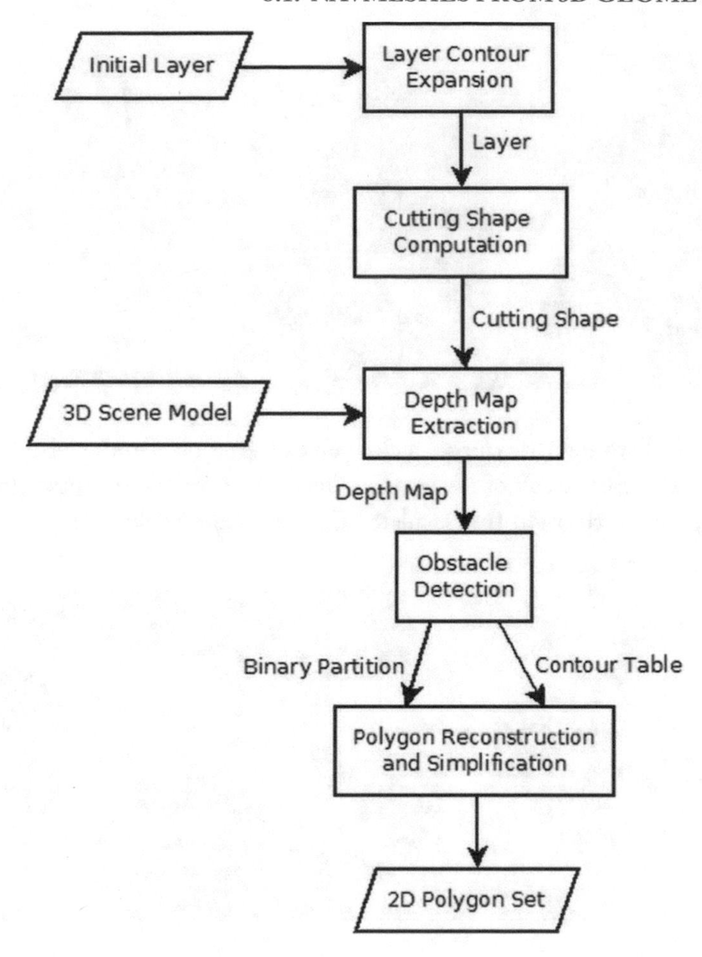

Figure 8.5: Layer refinement diagram.

3. $\cos(\alpha_{max}) > (\vec{n} \cdot \vec{UP})$. This means that the current surface, with normal $\vec{n}$ cannot be traversed by the character due to the given maximum walkable slope angle α_{max}.

In any other case, the fragment is passed to the next step of the graphics pipeline. Since the CS can intersect with vertical walls, and those walls will not produce any fragment in the rasterization process, we need to deal with near-perpendicular polygons separately in order not to lose any details of the original geometry. This is done by storing those polygons, and then for each layer rendering polygons that intersect with *accessible* voxels directly onto the depth map of the corresponding layer. Figure 8.7 shows the result of this process for the bottom and top layer of the example scene. Black areas indicate non-walkable space (obstacles, or empty) and grays denote the depth of the walkable areas. Notice how the ramp has been split between the two layers.

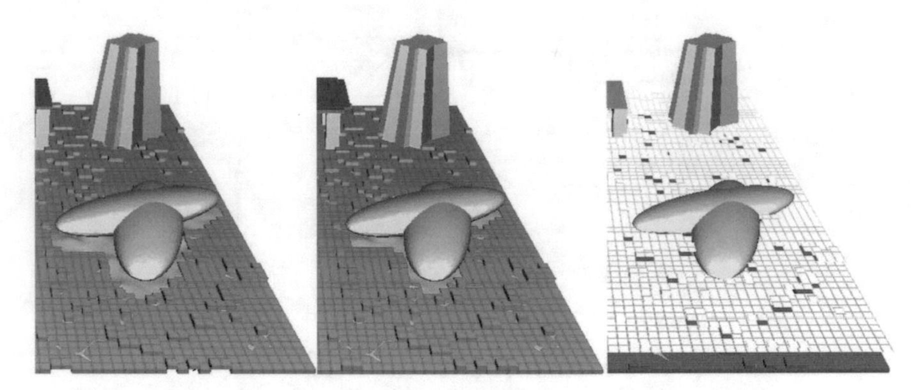

Figure 8.6: From left to right we can see a close up of the scene. On the left, the output of the layer extraction, in the center the result of the voxel expansion step to better capture the original geometry around obstacles, and on the right the calculated *Cutting Shape* (in white).

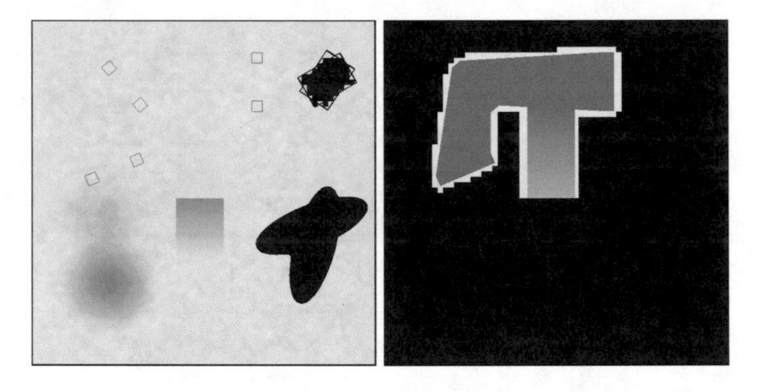

Figure 8.7: Depth maps for the bottom and top layers of the example scene.

Obstacle detection and polygon reconstruction

To differentiate floors from obstacles, a flood fill algorithm is performed over the depth map. Obstacles are detected when the difference in height between neighboring pixels is larger than the character's step height, h_s. The result is a binary file where 1 means floor, and 0 means obstacle. Pixels belonging to the contour of an obstacle are considered vertices of the obstacle shape. To reduce the final number of vertices, we apply the Ramer-Douglas-Peucker algorithm [48].

Note that we are reconstructing the shape of each polygon from a depth map, therefore the resolution used to generate this map will have an impact on the quality of the reconstructed shapes. Fine resolution will provide close alignment with the original geometry, albeit with a higher computational cost. In contrast, coarse resolution will provide speed at the cost of worse alignment with the original geometry (i.e., vertices of the polygons may be slightly displaced).

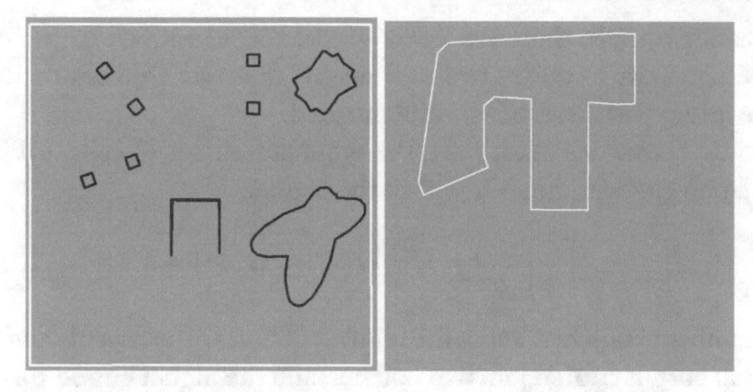

Figure 8.8: Floor plans of bottom and top layers. White polygons provide the shape of each layer and red polygons represent obstacles.

8.1.4 NAVMESH GENERATION

The final step is to generate a navigation mesh from the separated layer and 2D polygon floor plan. The outer polygon represents the contour of the floor, and the holes represent the obstacles. Each floor plan is used as an input to the navigation mesh generator, ANavMG [199], which calculates a near-optimal convex decomposition that will be used as a cell and portal graph (CPG). ANavMG calculates for every notch (concave vertex) its *Area of Interest* (delimited by the prolongation of the edges incident to the notch). Then the closest element to the notch inside this *Area of Interest* is calculated and a portal is created to convert the notch into a convex vertex. The closest element to the notch can be another vertex, an edge of the geometry, or a previously created portal. Figure 8.9 (left) shows the output provided by ANavMG for a simple polygon with holes.

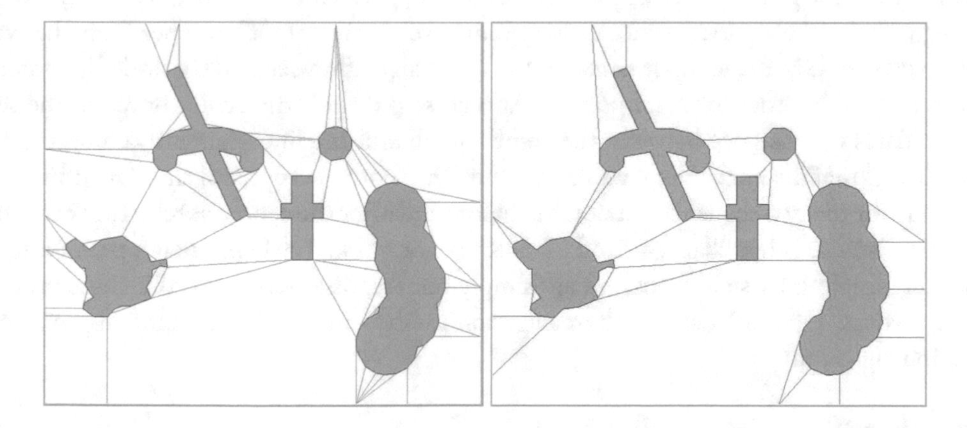

Figure 8.9: On the left we can see a Cell-and-Portal Graph obtained with ANavMG (69 cells), on the right we can see the CPG after applying convexity relaxation ($\tau_{cr} = 0.5$). The final number of cells is 29.

When creating portals, there are a few special considerations. If the closest element is a portal, it may be necessary to create two new portals to avoid T-Joints between portals. When this happens the initial portal can often be eliminated.

As proven by Fernandez et al. [62], the optimal number of cells, OPT, for a convex decomposition of a polygon with holes is within the bounds:

$$\left\lceil \frac{r}{2} \right\rceil + 1 - h \leq OPT \leq 2r + 1 - h$$

Where r is the number of notches, and h the number of holes. The lower bound corresponds to the ideal case where all portals created join two notches, and the higher bound corresponds to the case where a portal needs to be created for each notch to turn it into a convex vertex. Since ANavMG creates in most cases one portal per notch, our convex decomposition will theoretically provide a result around $r - h$, although in practice as we increase the complexity of the polygons our result tends toward the lower bound. Since our method provides a decomposition that is always guaranteed to be within the optimal bound, and our experimental results show a tendency toward the lower values of this bound, we can determine that our algorithm provides a near-optimal decomposition.

Convexity Relaxation

Local movement algorithms for obstacle avoidance may pay inordinate attention to small concavities in walls. It is beneficial to further reduce the number of cells in the final navigation mesh by slightly relaxing the notion of convexity, set empirically as a convexity relaxation threshold, τ_{cr}. This approach is derived from the Ramer-Douglas-Peucker algorithm [48], which reduces the number of points in a curve that is approximated by a series of points. Our method focuses exclusively on notches in the floor plan, since some of these minor notches can be ignored when creating portals. The algorithm checks every sequence of one or more notches found between two convex vertices. Given the input threshold τ_{cr}, the algorithm recursively finds the notches that need to be kept in order to create portals. At each step it finds the center notch of the sequence and calculates the distance between the center notch and the line segment joining the first and last vertex. If that distance is larger than τ_{cr} then that notch is kept and the algorithm calls itself recursively for the two sequences of notches before and after the center notch. The recursion stops when the distance is less than τ_{cr}, and all notches not marked as kept are ignored. Note that we are not eliminating these notches, we are simply not creating portals to split these notches into convex vertices. Figure 8.9 shows the navigation mesh before and after applying this convexity relaxation algorithm.

Merging layers

During the layer extraction step, the red channel (R) of the CS texture stores the value 0.5 when a pixel could belong to a boundary between layers. This information is used to determine the portals that join a layer with its neighboring layers.

Next, for each boundary pixel, find the two closest vertices between neigbouring layers and merge them into a single vertex. The position of the merged vertex is calculated as the average between the two positions of the vertices (note that the resolution of the depth map could lead to numerical differences). This can have the undesirable effect of adding an extra notch to the map. However this effect can be easily mitigated by the convexity relaxation method.

The algorithm iterates through these possible portal edges to join them together. After the merging process completes, the navigation mesh could end up with some non-essential portals. A non-essential portal is defined as a portal that if removed will not leave a notch in the navigation mesh. For example in Fig. 8.10 we see that a portal has been created in the center of the ramp to join the bottom and top layers (black dotted line) that could be removed merging the cells of the ramp. Therefore the final step of the algorithm consists of searching for non-essential portals around the merged areas.

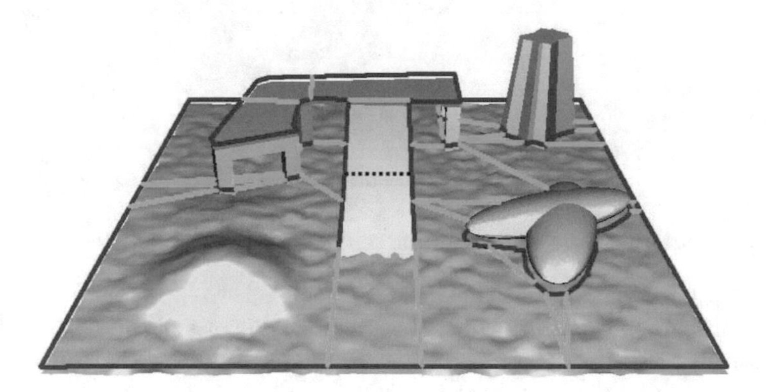

Figure 8.10: The merging step will join the top and bottom Navigation Meshes into one, and will also eliminate the non-essential portals around merged areas (black dotted line).

8.2 RESULTS

We have applied this algorithm to several scenarios of increasing complexity and number of vertices. NEOGEN has successfully generated the NavMesh for all the multi-layered 3D environments tested (Table 8.1). Figures 8.13 and 8.14 show the visual results obtained. Notice that our algorithm is robust in spite of intersecting geometry, cracks, and holes (which would be treated as obstacles).

Figure 8.11 shows the time taken by NEOGEN to output the NavMesh for each scene (tested with Intel Core 2 Quad Q9300 @ 2.50GHz, 4GB of RAM, and a GeForce 460 GTX). Even though the execution time of the algorithm is not a major goal, it is important that the process run as fast as possible, to make it easier for the designer to make changes to the geometry and see the impact on the resulting navigation mesh at interactive rates.

Table 8.1: Summary of the scenes tested (with references to the figure number) with the number of triangles and layers for each scene and the final number of cells generated by NEOGEN for the given convexity relaxation threshold, τ_{cr}

scene	#Fig	#triangles	#layers	#cells	τ_{cr}
map1	8.1	18,431	2	29	0.5
map2	8.14	7,308	3	86	0.5
map3	8.13	19,510	4	50	0.75

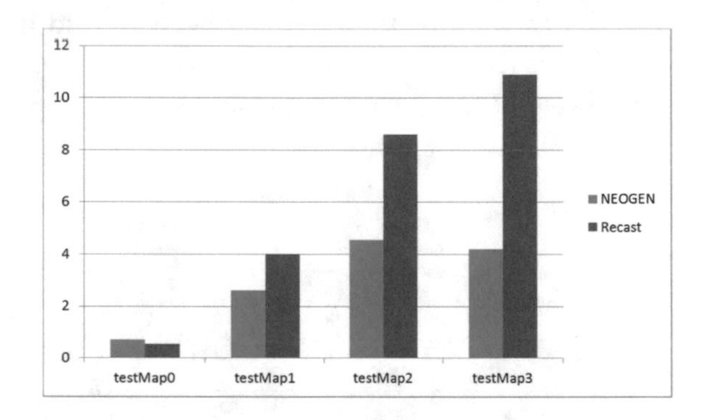

Figure 8.11: Computation time taken for each of the tested scenes.

We also compare our results against Recast, since it is one of the most widely used tools for the computation of NavMeshes in complex virtual applications. NEOGEN takes considerably less time to compute the NavMesh, and this difference increases with the size and complexity of the environment. For environment map1, Recast created 121 cells, whereas NEOGEN needed only 53 (without convexity relaxation) and can be further reduced to 29 (with $\tau_{cr} = 0.5$).

Another advantage to NEOGEN is that it creates cells that fit tightly to the geometry. Figure 8.12 compares the adjustment to the contour offered by NEOGEN and Recast. NEOGEN provides a better fit to the contour of the obstacles, and hence, its description of the walkable space is more accurate, e.g., in the contour around the columns. (Notice that the small visual offset between the cell limits and the contours is because the cell edges are rendered slightly above the terrain.) The voxelization used in Recast provides a coarse approximation of the geometry that can be refined by reducing the size of the voxels, but at the cost of generating significantly more cells.

Recast may create walkable regions that a character cannot access (Fig. 8.13), because it does not take into account any connectivity. Some of those regions can be discarded if they are small enough compared to the actual walkable region, but if not the resulting CPG has many unnecessary cells.

These results show that NEOGEN is faster than other NavMesh generation options, and yet requires minimal user input. The information required from the user could be limited to the

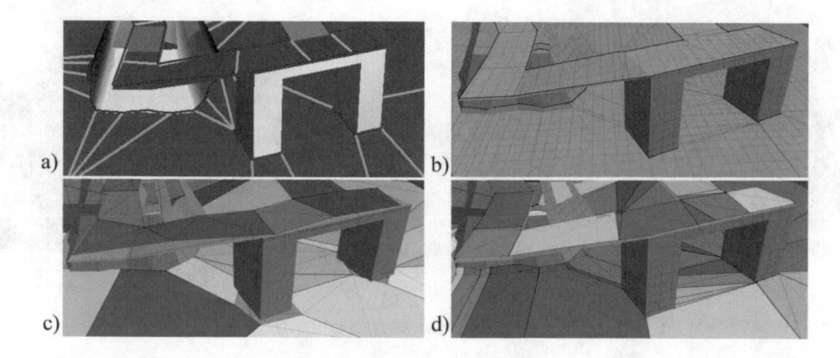

Figure 8.12: Adjustment to the geometry of the NavMesh in our method (a) and Recast (b,c,d). Recast results are calculated with the parameter *agents radius* set to 0 (to eliminate the object enlargement carried out by Recast to account for clearance). In (b) we can observe a good adjustment to the geometry when the parameter *cell size* used for the voxelization is set to 0.1, which has the drawback of creating too many cells (see (d) where cells are shown with different colors), and taking too long to compute. When the *cell size* is set to a large value (0.9), Recast can obtain a small number of cells (c), but with very poor adjustment to the geometry and even intersections (notice the bottom of the columns). In our result (a) we combine a small number of cells with tight adjustment (cell borders drawn slightly above the geometry for clarity).

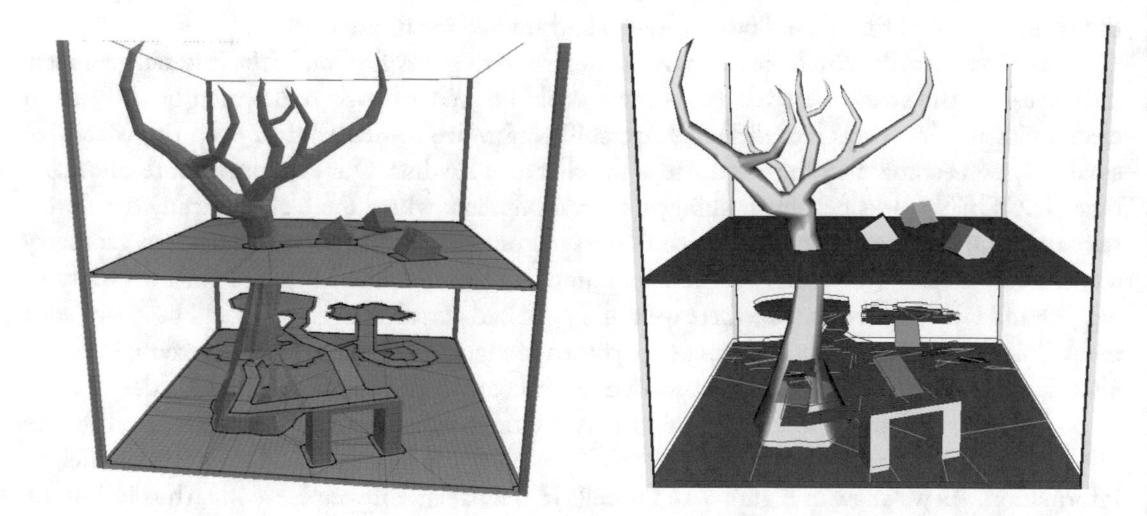

Figure 8.13: Comparison between the results given by Recast and NEOGEN (using map3). NEOGEN not only successfully eliminates unreachable layers, but also calculates a NavMesh with significantly fewer cells.

Figure 8.14: Two views of the NavMesh for the scene map2.

walkable seed s_w, since the other inputs (h_c, h_s, α_{max} and τ_{cr}) could be automatically extracted from the character and motion datasets.

8.3 LIMITATIONS AND DISCUSSION

NEOGEN provides an automatic algorithm to compute a navigation mesh from a given 3D geometry. The algorithm is capable of correctly splitting the environment into layers, creating a navigation mesh per layer and then linking them through portals. The main advantage of NEOGEN is to provide a tight adjustment to the input geometry while keeping the total number of cells as close as possible to the optimal subdivision. However, there are certain limitations of the current method that should be considered and studied for future work.

For instance, the voxel size can have an impact on the classification of the original geometry into layers. If the voxels were too large, there could be scenarios where it would be difficult to clearly classify the voxels into different layers. Therefore we recommend to keep the voxel size small. A good estimate for the voxel size is the character's radius. Our experimental results show that NEOGEN works best for building-like environments where there are typically flat floors, stair and ramps. It also works well with outdoor environments as long as the changes in geometry are not very drastic (too many consecutive bumps and non-smooth surfaces with a variety of heights and layers). An example where we could get a bad partition in layers would be a cave with many holes, stalactites, or stalagmites as the shown in Figure 8.15. This particular example would be difficult to capture by our voxelization due to the coarse representation in the Y axis.

It is also important to keep in mind that the tight adjustment is only guaranteed on the 2D projection of each layer, which means that the polygons reconstructed do not store height information. As we show in Figure 8.16 the cells in the navigation graph are 2D, thus losing the information regarding the bumps in the terrain. However, for the purpose of having characters moving through this navigation meshes, height fields can be used to store the elevation of the terrain.

Finally, the resolution of the depth map will have an impact on the adjustment of the reconstructed polygons to the original geometry. In Figures 8.16 and 8.17 we can see examples

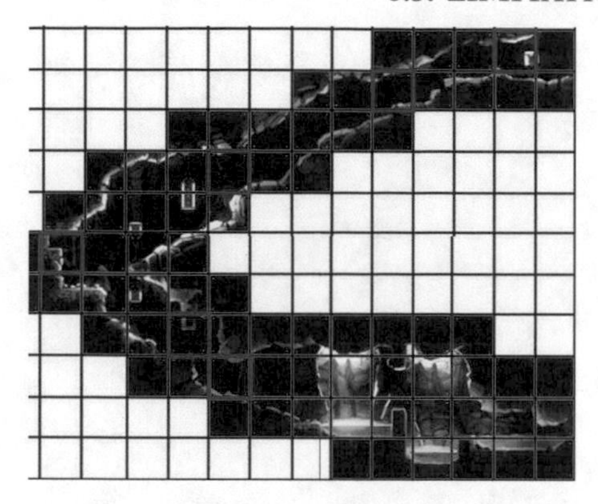

Figure 8.15: Example of complex cave-like environment that would be difficult to process due to the lower resolution used in the Y axis. The grid shows an example of voxelization, where voxels with both ground and ceiling geometry have been highlighted in red.

Figure 8.16: Examples of cases where a coarse depth map can lead to extra notches appearing around obstacles during polygon reconstruction. Another undesirable outcome may be polygons that are not an exact adjustment to the original geometry.

of the consequences of using a coarse resolution. The figures show how a coarse resolution may introduce extra notches in the polygon reconstruction phase.

In general, if we choose a high resolution and a small voxel size (based on character's features) we can obtain a high quality navigation mesh. The disadvantage of using high resolution is mainly the computational time, which should not be a problem as usually navigation meshes are calculated off line and not modified during simulation. However if dynamic navigation meshes were needed, then this would be an important trade off.

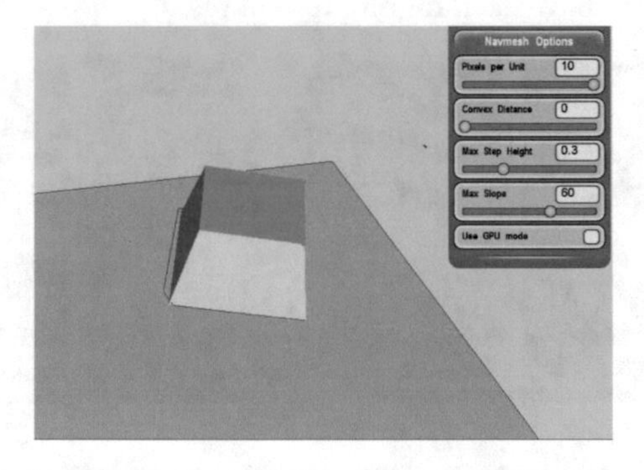

Figure 8.17: Polygon reconstruction over the original geometry, where we can see another example of poor adjustment due to a coarse depth map.

CHAPTER 9

Multi-domain Planning in Dynamic Environments

Mubbasir Kapadia, Alejandro Beacco, Francisco Garcia, Vivek Reddy, Nuria Pelechano, and Norman I. Badler

9.1 MULTI-DOMAIN PLANNING

Now that we can readily compute *where* an agent can step, the next task is to decide *how* to achieve its destination. Contemporary interactive applications require high fidelity navigation of interacting agents in non-deterministic, dynamic virtual worlds. The environment and agents are constantly affected by unpredictable forces (e.g., human input), making it impossible to accurately extrapolate the future world state to make optimal decisions. These complex domains require robust navigation algorithms that can handle partial and imperfect knowledge, while still making decisions which satisfy space-time constraints.

Different situations require different granularity of control. An open environment with no other agents and only static obstacles requires only coarse-grained control. Cluttered dynamic environments require fine-grained character control with carefully planned decisions with spatial *and* temporal precision. Some situations (e.g., potential deadlocks) may require explicit coordination between multiple agents.

The problem domain of interacting agents in dynamic environments is extremely high-dimensional and continuous, with infinite ways to interact with objects and other agents. Having a rich action set, and a system that makes intelligent action choices, facilitates robust, realistic

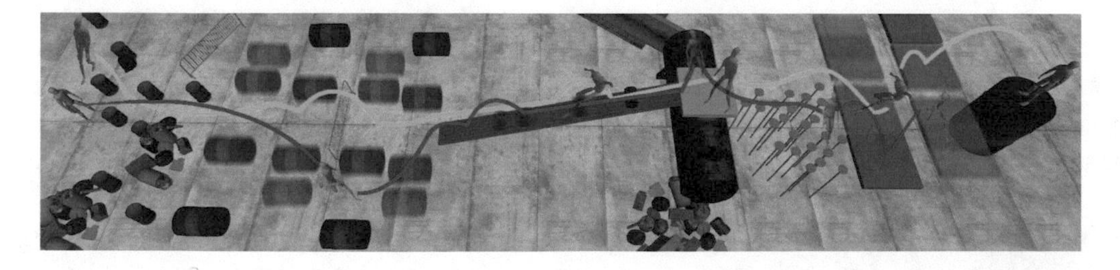

Figure 9.1: Two agents navigating with space-time precision through a complex dynamic environment.

virtual characters, at the expense of interactivity and scalability. Greatly simplifying the problem domain yields interactive virtual worlds with hundreds and thousands of agents that exhibit only simple behaviors. The ultimate, far-reaching goal is still a considerable challenge: a real-time system for autonomous character control that can handle many characters, without compromising control fidelity.

Typical agent simulations decouple global navigation [112, 268] and local collision avoidance [208], or demonstrate real-time space-time planning for constrained global navigation using a single character [160]. These approaches necessitate trade-offs between number of agents, control fidelity, and environment complexity. So far, no previously proposed technique efficiently accounts for the dynamic nature of the environment at all levels of the decision-making process.

To advance real-time multi-character navigation planning, we construct a framework that uses multiple heterogeneous problem domains of differing complexities for navigation in large, complex, dynamic virtual environments. These problem domains (spaces of decision-making) differ in the complexity of their state representations and the fidelity of agent control. Domains range from a static navigation mesh which only accounts for static objects in the environment, to a space-time domain that factors in dynamic obstacles and other agents at much finer resolution. These domains provide different trade-offs in performance and fidelity of control, requiring a framework that efficiently works in multiple domains by using plans in one domain to focus and accelerate searches in more complex domains.

A global planning problem (start and goal configuration) is dynamically decomposed into a set of smaller problem instances across different domains, where an anytime dynamic planner is used to efficiently compute and repair plans for each of these problems. Planning tasks are connected by either using the computed path from one domain to define a *tunnel* to focus searches, or using successive waypoints along the path as start and goal for a planning task in another domain to reduce the search depth, thereby accelerating searches in more complex domains. Using this framework, we demonstrate real-time character navigation for multiple agents in large-scale, complex, dynamic environments, with precise control, and little computational overhead. The related publication can be found here [119].

9.2 OVERVIEW

The problem domain of a planner determines its effectiveness in solving a particular problem instance. A complex domain that accounts for many factors such as dynamic environments and other agents, and has a large branching factor in its action space, can solve more difficult problems, but at a larger cost. A simpler domain benefits computational efficiency while compromising on control fidelity. Our framework enables the use of multiple heterogeneous domains of control, providing a balance between control fidelity and computational efficiency, without compromising either.

A global problem instance P_0 is dynamically decomposed into a set of smaller problem instances $\{P^{'}\}$ across different planning domains $\{\Sigma_i\}$. Section 9.3 describes the different domains, and Section 9.4 describes the problem decomposition across domains. Each problem instance P' is assigned a planning task $T(P')$, and an anytime dynamic planner is used to efficiently compute and repair plans for each of these tasks, while using plans in one domain to focus and accelerate searches in more complex domains. Plan efforts across domains are reused in two ways. The computed path from one domain can be used to define a *tunnel* which focuses the search, reducing its effective branching factor. Each pair of successive waypoints along a path can also be used as (start,goal) pairs for a planning task in another domain, thus reducing the search depth. Both these methods are used to focus and accelerate searches in complex domains, providing real-time efficiency without compromising on control fidelity. Section 9.5 describes the relationships between domains.

9.3 PLANNING DOMAINS

A problem domain is defined as $\Sigma = \langle \mathbb{S}, \mathbb{A}, \mathtt{c}(s,s'), \mathtt{h}(s, s_{goal}) \rangle$, where the state space $\mathbb{S} = \{\mathbb{S}_{self} \times \mathbb{S}_{env} \times \mathbb{S}_{agents}\}$ includes the internal state of the agent $\mathbb{S}_{self}$, the representation of the environment $\mathbb{S}_{env}$, and other agents $\mathbb{S}_{agents}$. $\mathbb{S}_{self}$ may be modeled as a simple particle with a collision radius. $\mathbb{S}_{env}$ can be an environment triangulation with only static information or a uniform grid representation with dynamic obstacles. $\mathbb{S}_{agents}$ is defined by the vicinity within which neighboring agents are considered. Imminent threats may be considered individually or just represented as a density distribution at far-away distances. The action space $\mathbb{A}$ defines the set of all possible successors $\mathtt{succ}(s)$ and predecessors $\mathtt{pred}(s)$ at each state s, as shown in Equation 9.1. Here, $\delta(s,i)$ describes the ith transition, and $\Phi(s,s')$ is used to check if the transition from s to s' is possible. The cost function $\mathtt{c}(s,s')$ defines the cost of transition from s to s'. The heuristic function $\mathtt{h}(s, s_{goal})$ defines the estimate cost of reaching a goal state.

$$\mathtt{succ}(s) = \{s + \delta(s,i) | \Phi(s,s') = \mathtt{TRUE} \ \forall i\} \tag{9.1}$$

A problem definition $P = \langle \Sigma, s_{start}, s_{goal} \rangle$ describes the initial configuration of the agent, the environment and other agents, along with the desired goal configuration in a particular domain. Given a problem definition P for domain Σ, a planner searches for a sequence of transitions to generate a space-time plan $\Pi(\Sigma, s_{start}, s_{goal}) = \{s_i | s_i \in \mathbb{S}(\Sigma)\}$ that takes an agent from s_{start} to s_{goal}.

9.3.1 MULTIPLE DOMAINS OF CONTROL

We define four domains which provide a nice balance between global static navigation and fine-grained space-time control of agents in dynamic environments. Figure 9.2 illustrates the different domain representations for a given environment.

Static Navigation Mesh Domain Σ_1. This domain uses a triangulated representation of free space and only considers static immovable geometry. Dynamic obstacles and agents are not considered in this domain. The agent is modeled as a point mass, and valid transitions are between connected

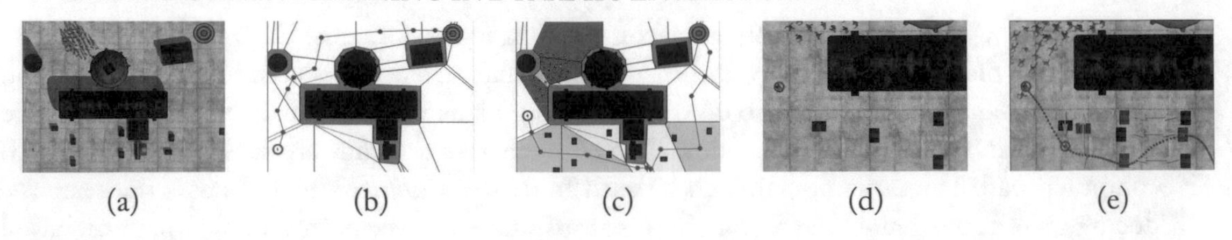

Figure 9.2: (a) Problem definition with initial configuration of agent and environment. (b) Global plan in static navigation mesh domain Σ_1 accounting for only static geometry. (c) Global plan in dynamic navigation mesh domain Σ_2 accounting for cumulative effect of dynamic objects. (d) Grid plan in Σ_3. (e) Space-time plan in Σ_4 that avoids dynamic threats and other agents.

free spaces, represented as polygons. The cost function is the straight line distance between the center points of two free spaces. Additional connections are also precomputed (or manually annotated) to represent transitions such as jumping with a higher defined cost. The heuristic function is the Euclidean distance between a state and the goal. Searching for an optimal solution in this domain is very efficient and quickly provides a global path for the agent to navigate.

Dynamic Navigation Mesh Domain Σ_2. This also uses triangulations to represent free spaces and coarsely accounts for dynamic properties of the environment to make a more informed decision at the global planning layer. The work in [296] embeds population density information in environment triangulations to account for the movement of agents at the global planning layer. We adopt a similar method by defining a time-varying density field $\phi(t)$ which stores the density of movable objects (agents and obstacles) for each polygon in the triangulation at some point of time t. $\phi(t_0)$ represents the density of agents and obstacles currently present in the polygon. The presence of objects and agents in polygons at future timesteps can be estimated by querying their plans (if available). The space-time positions of deterministic objects can be accurately queried while the future positions of agents can be approximated based on their current computed paths, assuming that they travel with constant speed along their path without deviation. $\phi(t)$ contributes to the cost of selecting a waypoint in Σ_2 during planning. The resolution of the triangulation may be kept finer than Σ_1 to increase the resolution of the dynamic information in this domain. Consequently a set of global waypoints are chosen in this domain which avoids crowded areas or other high cost regions.

Grid Domain Σ_3. The grid domain discretizes the environment into grid cells where a valid transition is considered between adjacent cells that are free (diagonal movement is allowed). An agent is modeled as a point with a radius—its orientation and speed is not considered in this domain. This domain only accounts for the current position of dynamic obstacles and agents, and cannot predict collisions in space-time. The cost and heuristic are functions that measure the Euclidean distance between grid cells.

Space-Time Domain Σ_4. This domain models the current state of an agent as a space-time position with a current velocity $(\mathbf{x}, \mathbf{v}, t)$. The figure illustrates schematically the state and action space in Σ_4, showing both a valid transition and an invalid transition due to a space-time collision with a neighboring agent. The transition function $\delta(s, i)$ for Σ_4 is defined below:

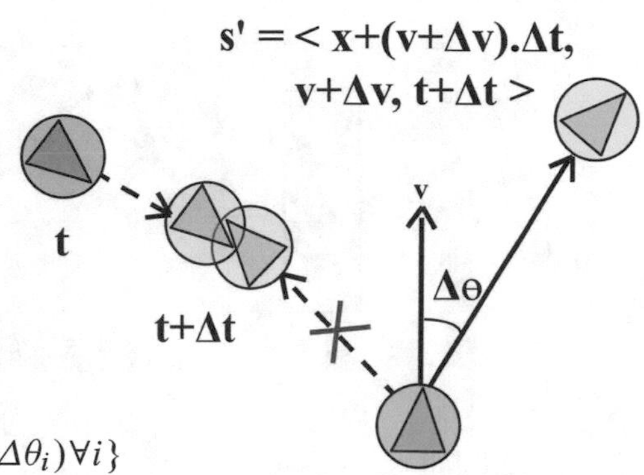

$$\delta(s, i) = \{\Delta \mathbf{v}_i \cdot \Delta t | \Delta \mathbf{v}_i = (\Delta v_i \cdot \sin \Delta\theta_i, \Delta v_i \cdot \cos \Delta\theta_i) \forall i\}$$

where $\Delta v = \{0, \pm a\}$ is the possible speed change and $\Delta\theta = \{0, \pm\frac{\pi}{8}, \pm\frac{\pi}{4}, \pm\frac{\pi}{2}\}$ is the possible orientation change the agent can make from its current state. For example, $\Delta \mathbf{v} = a, \Delta\theta = \frac{\pi}{8}$ produces a transition where the agent accelerates by a for the duration of the timestep and rotates by $\frac{\pi}{8}$. The bounds of $\Delta\theta$ are limited between $\{-\frac{\pi}{2}, \frac{\pi}{2}\}$ to constrain the maximum rate of turning. Transitions are also bound so that the speed and acceleration of an agent cannot exceed a given threshold. Jumps are modeled as a high cost transition between two space-time points such that the region between them may be occupied or untraversable for that time interval. Despite the coarse discretization of $\Delta\theta$, the branching factor of this domain is much higher, providing a greater degree of control fidelity with added computational overhead.

Σ_4 accounts for all obstacles (static and dynamic) and other agents. The traversability of a grid cell is queried in space-time by checking to see if movable obstacles and agents occupy that cell at that particular point of time, by using their published paths. For space-time collision checks, only agents and obstacles are considered that are within a certain region from the agent, defined using a foveal angle intersection. The cost and heuristic definitions have a great impact on the performance in Σ_4. We use an energy-based cost formulation that penalizes change in velocity with a non-zero cost for zero velocity. Jump transitions incur a higher cost. The heuristic function penalizes states that are far away from s_{goal} in both space and time. This is achieved using a weighted combination of a distance metric and a penalty for a deviation of the current speed from the speed estimate required to reach s_{goal}.

The domains described here are *not* a comprehensive set and only serve to showcase the ability of our framework to use multiple heterogeneous domains of control in order to solve difficult problem instances at low computation cost. Our framework can be easily extended to use other domain definitions (e.g., a footstep domain), as described in Chapter 3.

9.4 PROBLEM DECOMPOSITION AND MULTI-DOMAIN PLANNING

Figure 9.4(a) illustrates the use of tunnels to connect each of the four domains, ensuring that a complete path from the agents initial position to its global target is computed at all levels.

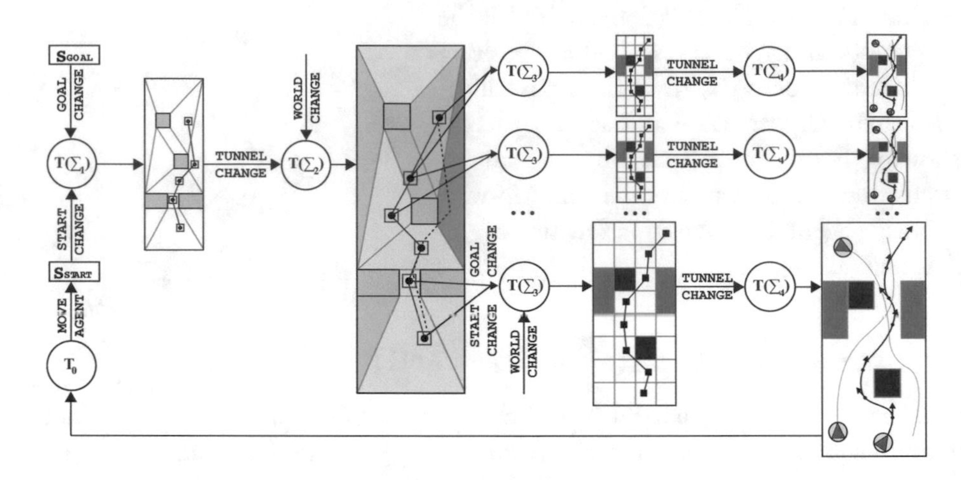

Figure 9.3: Expanded illustration of domain relationship shown in Figure 9.4(b). A global problem instance (start and goal state) is decomposed into a set of smaller problem instances across multiple planning domains. Planning tasks $T(\Sigma)$ are assigned to each of these problems and scheduled using a dynamic priority scheme based on events from the environment and other tasks.

Figure 9.4(b) shows how Σ_2 and Σ_3 are connected by using successive waypoints in $\Pi(\Sigma_2)$ as start and goal for independent planning tasks in Σ_3. This relation between Σ_2 and Σ_3 allows finer-resolution plans to be computed between waypoints in an independent fashion. Limiting Σ_3 (and Σ_4) to plan between waypoints instead of the global problem instance insures that the search horizon in these domains is never too large, and that fine-grained space-time trajectories to the initial waypoints are computed quickly. However, completeness and optimality guarantees are relaxed as Σ_3, Σ_4 never compute a single path to the global target.

Figure 9.3 illustrates the different events that are sent between planning tasks to trigger plan refinement and updates for the domain relationship in Fig. 9.4(b). Σ_1 is first used to compute a path from s_{start} to s_{goal}, ignoring dynamic obstacles and other agents. $\Pi(\Sigma_1)$ is used to accelerate computations in Σ_2, which refines the global path to factor in the distribution of dynamic objects in the environment. Depending on the relationship between Σ_2 and Σ_3, a single planning task or multiple independent planning tasks are used in Σ_3. Finally, the plan(s) of $T(\Sigma_3)$ are used to accelerate searches in Σ_4.

Changes in s_{start} and s_{goal} trigger plan updates in $T(\Sigma_1)$, which are propagated through the task dependency chain. $T(\Sigma_2)$ monitors plan changes in $T(\Sigma_1)$ as well as the cumulative effect of changes in the environment to refine its path. Each $T(\Sigma_3)$ instance monitors changes in the waypoints along $\Pi(\Sigma_2)$ to repair its solution, as well as nearby changes in obstacle and agent position. Finally, $T(\Sigma_4)$ monitors plan changes in $T(\Sigma_3)$ (which it depends on) and repairs

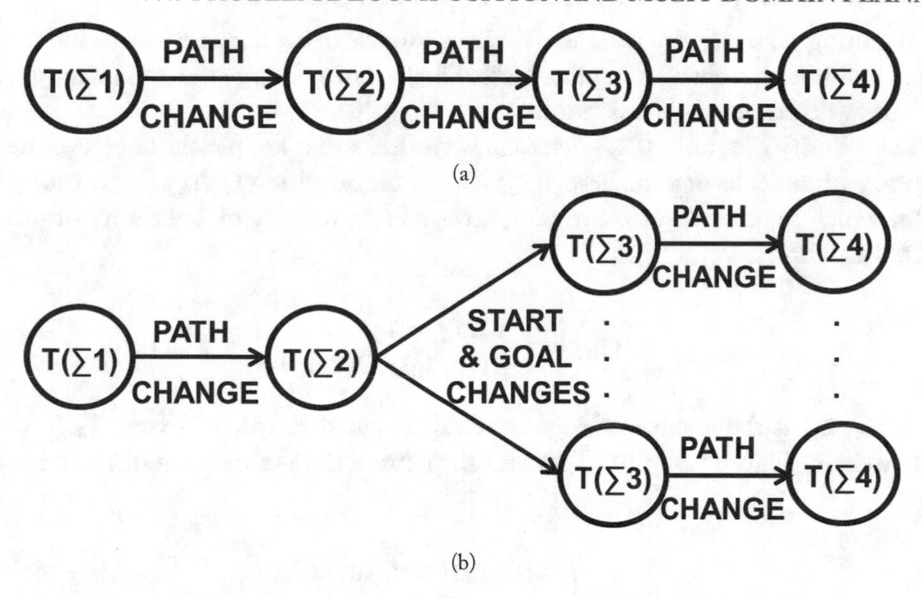

Figure 9.4: Relationship between domains. (a) Use of tunnels to connect each of the four domains. (b) Use of successive waypoints in $\Pi(\Sigma_2)$ as start, goal pairs to instantiate multiple planning tasks in Σ_3 and Σ_4.

its solution to compute a space-time trajectory that avoids collisions with static and dynamic obstacles and other agents.

Events are triggered (outgoing edges) and monitored (incoming edges) by tasks, creating a cyclic dependency between tasks, with $\mathtt{T}_0$ (agent execution) monitoring changes in the plan produced by the particular $T(\Sigma_4)$, which monitors the agent's most imminent global waypoint. Tasks that directly affect the agent's next decision, and tasks with currently invalid or sub-optimal solutions are given higher priority. Given the maximum amount of time to deliberate t_{max}, the agent pops one or more tasks that have highest priority and divides the deliberation time across tasks (most imminent tasks are allocated more time). Task priorities constantly change based on events triggered by the environment and other tasks.

9.4.1 PLANNING TASKS AND EVENTS

A task $T(P)$ is a planner which is responsible for generating and maintaining a valid (and ideally optimal) solution for a particular problem definition $P = \langle \Sigma, s_{start}, s_{goal} \rangle$ where s_{start}, s_{goal}, and the search graph may be constantly changing. There are four types of tasks, each of which solves a particular problem in the domains described above. An additional task $\mathtt{T}_0$ is responsible for moving the agent along the path, while enforcing steering and collision constraints.

Events are triggered and monitored by planning tasks in different domains, as illustrated in Figure 9.3. Changes in start and goal, or environment changes may potentially invalidate current plans, requiring plan refinement. Tasks that use tunnels to accelerate searches in more complex domains monitor plan changes in other tasks. Finally, tasks observe the optimality status of their own plans to determine their task priority.

The priority of a task $p(\mathtt{T}_a)$ determines the tasks that are picked to be executed at every time step, with tasks having smallest $p(\mathtt{T}_a)$ chosen for execution ($p(\mathtt{T}_a)$ is short for $p(T(\Sigma_a))$). Task $\mathtt{T}_0$, which handles agent movement, always has a priority of 1. Priority of other tasks is calculated as follows:

$$p(\mathtt{T}_a) = \begin{cases} 1 \text{ if } \mathtt{T}_a = \mathtt{T}_0 \\ \mu(\mathtt{T}_a, \mathtt{T}_0) \cdot \Omega(\mathtt{T}_a) \text{ else} \end{cases} \tag{9.2}$$

where $\mu(\mathtt{T}_a, \mathtt{T}_0)$ is the number of edge traversals required to reach $\mathtt{T}_0$ from $\mathtt{T}_a$ in the task dependency chain (Figure 9.3). $\Omega(\mathtt{T}_a)$ denotes the current state of the plan of $\mathtt{T}_a$ and is defined as follows:

$$\Omega(\mathtt{T}_a) = \begin{cases} 1 \text{ if } \mathtt{SOLUTION\_INVALID} \\ \epsilon \text{ if plan inflation factor, } \epsilon > 1 \\ \infty \text{ if plan inflation factor, } \epsilon = 1 \end{cases} \tag{9.3}$$

where ϵ is the inflation factor used to determine the optimality bounds of the current plan for that task. The agent pops one or more tasks that have highest priority and divides the deliberation time available across tasks, with execution-critical tasks receiving more time. Tasks that have the same priority are ordered based on task dependency. Hence, $\mathtt{T}_0$ is always executed at the end of every update after all planning tasks have completed.

The overall framework enforces strict time constraints. Given an allocated time to deliberate for each agent (computed based on desired frame rate and number of agents), the time resource is distributed based on task priority. In the remote event that there is no action to execute, the agent remains stationary (no impact on frame-rate) for a few frames (fraction of a second) until a valid plan is computed.

9.5 RELATIONSHIP BETWEEN DOMAINS

The complexity of the planning problem increases exponentially with increase in dimensionality of the search space—making the use of high-dimensional domains nearly prohibitive for real-time applications. In order to make this problem tractable, planning tasks must efficiently use plans in one domain to focus and accelerate searches in more complex domains. Section 9.5.1 describes a method for mapping a state from a low-dimensional domain to one or more states in a higher dimensional domain. The remainder of this section describes two ways in which plans in one domain can be used to focus and accelerate searches in another domain.

9.5.1 DOMAIN MAPPING

We define a $1:n$ function $\lambda(s, \Sigma, \Sigma')$ that allows maps states in $\mathbb{S}(\Sigma)$ to one or more equivalent states in $\mathbb{S}(\Sigma')$.

$$\lambda(s, \Sigma, \Sigma') : s \rightarrow \{s' | s' \in \mathbb{S}(\Sigma') \wedge s \equiv s'\} \tag{9.4}$$

The mapping functions are defined specifically for each domain pair. For example, $\lambda(s, \Sigma_1, \Sigma_2)$ maps a polygon $s \in \mathbb{S}(\Sigma_1)$ to one or more polygons $\{s' | s' \in \mathbb{S}(\Sigma_2)\}$ such that s' is spatially contained in s. If the same triangulation is used for both Σ_1 and Σ_2, then there exists a one-to-one mapping between states. Similarly, $\lambda(s, \Sigma_2, \Sigma_3)$ maps a polygon $s \in \mathbb{S}(\Sigma_2)$ to multiple grid cells $\{s' | s' \in \mathbb{S}(\Sigma_3)\}$ such that s' is spatially contained in s. $\lambda(s, \Sigma_3, \Sigma_4)$ is defined as follows:

$$\lambda(s, \Sigma_3, \Sigma_4) : (\mathbf{x}) \rightarrow \{(\mathbf{x} + W(\Delta\mathbf{x}), t + W(\Delta t))\} \tag{9.5}$$

where $W(\Delta)$ is a window function in the range $[-\Delta, +\Delta]$. The choice of t is important in mapping Σ_3 to Σ_4. Since we use λ to effectively map a plan $\Pi(\Sigma_3, s_{start}, s_{goal})$ in Σ_3 to a tunnel in Σ_4, we can exploit the path and the temporal constraints of s_{start} and s_{goal} to define t for all states along the path. The total path length and the time to reach s_{goal} are calculated. This yields the approximate time of reaching a state along the path, assuming the agent is traveling at a constant speed.

9.5.2 MAPPING SUCCESSIVE WAYPOINTS TO INDEPENDENT PLANNING TASKS

Successive waypoints along the plan from one domain can be used as start and goal for a planning task in another domain. This effectively decomposes a planning problem into multiple independent planning tasks, each with a much smaller search depth.

Consider a path $\Pi(\Sigma_2) = \{s_i | s_i \in \mathbb{S}(\Sigma_2), \forall i \in (0, n)\}$ of length n. For each successive waypoint pair (s_i, s_{i+1}), we define a planning problem $P_i = \langle \Sigma_3, s_{start}, s_{goal} \rangle$ such that $s_{start} = \lambda(s_i, \Sigma_2, \Sigma_3)$ and $s_{goal} = \lambda(s_{i+1}, \Sigma_2, \Sigma_3)$. Even though λ may return multiple equivalent states, we choose only one candidate state. For each problem definition P_i, we instantiate an independent planning task $T(P_i)$ which computes and maintains path from s_i to s_{i+1} in Σ_3. Figure 9.4 illustrates this connection between Σ_2 and Σ_3.

9.6 RESULTS

9.6.1 COMPARATIVE EVALUATION OF DOMAIN RELATIONSHIPS

We randomly generate $1,000$ scenarios of size $100m \times 100m$, with random configurations of obstacles (both static and dynamic), start state, and goal state, then record the effective branching factor, number of nodes expanded, time to compute a plan, success rate, and quality of the plans obtained. The effective branching factor is the average number of successors that were generated over the course of one search. Success rate is the ratio of the number of scenarios for which a collision-free solution was obtained. Plan quality is the ratio of the length of the static optimal path and the path obtained. A plan quality of 1 indicates that the solution obtained was able to minimize distance without any deviations. Similar metrics for analyzing multi-agent simulations have been used in [134]. The aggregate metrics for the different domains and domain relationships are shown in Table 9.1. Rows 3 and 6 in Table 9.1 include the added time to compute plans in earlier domains for tunnel search, to provide an absolute basis of comparison. All experiments were performed on a single-threaded 2.80 GHz Intel(R) Core(TM) i7 CPU.

Σ_1 and Σ_2 can quickly generate solutions but are unable to solve most of the scenarios as they do not resolve fine-grained collisions. The use of plans from Σ_1 accelerates searches in Σ_2 (Table 9.1, Row 3). However, the real benefit of using both Σ_1 and Σ_2 is evident when performing repeated searches across domains in large environments when an initial plan $\Pi(\Sigma_1)$ accelerates repeated refinements in Σ_2 (and other subsequent domains). Using Σ_3 in a large environment takes much longer to produce similar paths. Σ_4 is unable to find a complete solution for large-scale problem instances (we limit maximum number of nodes expanded to 10^4), and the partial solutions often suffer from local minima, resulting in a low success rate. The benefit of using tunnels is evident in the dramatic reduction of the effective branching factor and nodes expanded for Σ_4.

When using the complete global path from Σ_3 as a tunnel for Σ_4 (Figure 9.4(a) and Row 6 in Table 9.1), the effective branching factor reduces from 21.5 to 5.6, producing an exponential drop in node expansion and computation time, and enabling complete solutions to be generated in the space-time domain. This planning task is able to successfully solve nearly 92% of the scenarios that were generated. However, since s_{start} and s_{goal} are far apart, the large depth of the search prevents this from being used at interactive rates for many agents.

By using successive waypoints in $\Pi(\Sigma_2)$ as s_{start} and s_{goal} to create a series of planning tasks in Σ_3 and Σ_4 (Fig. 9.4(b) and Row 7 in Table 9.1), we reduce the breadth *and* depth of the search, allowing solutions to be returned at a fraction of the time (6 ms), without affecting the success rate. The trade-off is that independent plans are generated between waypoints along the global path, creating a two-level hierarchy between the domains.

Conclusion. The comparative evaluations of domains shows that no single domain can efficiently solve the challenging problem instances that were sampled. The use of tunnels significantly reduces the effective branching factor of the search in Σ_3 and Σ_4, while mapping successive waypoints in $\Pi(\Sigma_2)$ to multiple independent planning tasks reduce the depth of the search in Σ_3 and Σ_4, without impacting success rate and quality. For the remaining results, we adopt this domain relationship as it works well for our application of simulating multiple goal-directed agents in dynamic environments at interactive rates. Users may choose a different relationship based on their specific needs.

Domain	BF	N	T	S	Q
$T(\Sigma_1)$	3.7	43	3	0.17	0.76
$T(\Sigma_2)$	4.6	85	8	0.23	0.57
$T(\Sigma_2, \Pi(\Sigma_1))$	2.1	17	5	0.32	0.65
$T(\Sigma_3)$	7.4	187	18	0.68	0.73
$T(\Sigma_4)$	21.5	10^4	2487	0.34	0.26
$T(\Sigma_4, \Pi(\Sigma_3, \Sigma_2, \Sigma_1))$	5.6	765	136	0.92	0.64
$\sum T_i(\Sigma_4, \Pi(\Sigma_3, \Sigma_2, \Sigma_1))$	5.4	75	8	0.86	0.58

Table 9.1: Comparative evaluation of the domains, and the use of multiple domains. **BF** = Effective branching factor. **N** = Average number of nodes expanded. **T** = Average time to compute plan (ms). **S** = Success rate of planner to produce collision-free trajectory. **Q** = Plan quality. Rows 6, 7 correspond to the domain relationships illustrated in Figs. 9.4(a) and (b), respectively

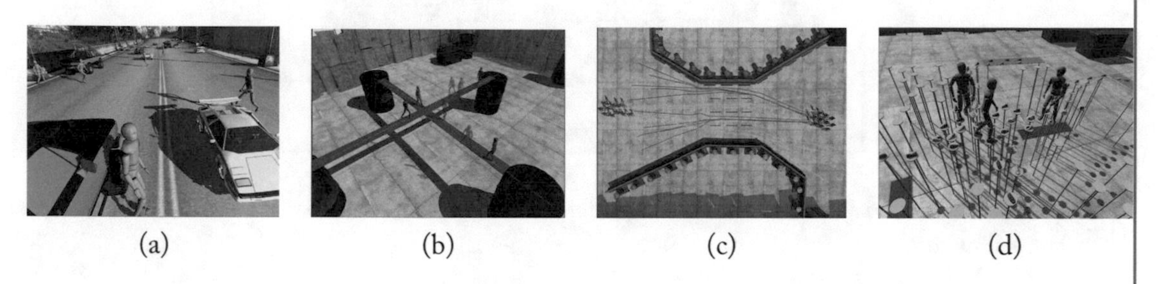

(a) (b) (c) (d)

Figure 9.5: Different scenarios. (a) Agents crossing a highway with fast moving vehicles in both directions. (b) 4 agents solving a deadlock situation at a 4-way intersection. (c) 20 agents distributing themselves evenly in a narrow passage, to form lanes both in directions. (d) A complex environment requiring careful foot placement to obtain a solution.

9.6.2 PERFORMANCE

We measure the performance of the framework by monitoring the execution time of each task type, with multiple instances of planning tasks for Σ_3 and Σ_4. We limit the maximum deliberation time $t_{max} = 10$ ms, which means that the total time executing any of the tasks at each frame cannot exceed $10ms$. For this experiment, we limit the total number of tasks that can be executed in a single frame to 2 (including $\mathtt{T}_0$) to visualize the execution time of each task over different frames. Figure 9.6 illustrates the task execution times of a single agent over a 30-second simulation for the scenario shown in Fig. 9.2(a). The execution task $\mathtt{T}_0$, which is responsible for character animation and simple steering, takes approximately 0.4 – 0.5 ms of execution time every frame. Spikes in the execution time correlate to events in the world. For example, a local non-deterministic change in the environment (frames 31,157) triggers a plan update in $T(\Sigma_3)$, which in turn triggers an update in $T(\Sigma_4)$. A global change such as a crowd blocking a passage or a change in goal (frames 39, 237,281) triggers an update in $T(\Sigma_2)$ or $T(\Sigma_1)$ which in turn propagates events down the task dependency chain.

Note that there are often instances during the simulation when the start or goal changes significantly or when plans are invalidated, requiring planning from scratch. However, we ensure

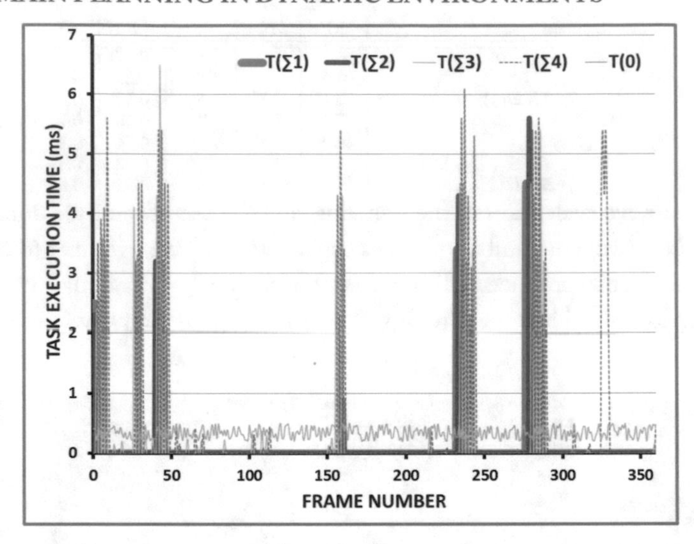

Figure 9.6: Task execution times of the different tasks in our framework over the course of a 60-second simulation.

that our framework meets real-time constraints due to the following design decisions: (a) limiting the maximum amount of time to deliberate for the planning tasks, (b) intelligently distributing the available computational resources between tasks with highest priority, and (c) increasing the inflation factor to quickly produce a sub-optimal solution when a plan is invalidated, and refining the plan in successive frames.

Memory. $T(\Sigma_1)$ and $T(\Sigma_2)$ precompute navigation meshes for the environment whose size depend on environment complexity, but are shared by all agents in the simulation. The runtime memory requirement of these tasks is negligible since it expands very few nodes. The memory footprint of $T(\Sigma_3)$ and $T(\Sigma_4)$ is defined by the number of nodes visited by the planning task during the course of a simulation. Since each planning task in Σ_3 and Σ_4 searches between successive waypoints in the global plan, the search horizon of the planners is never too large. On average, the number of visited nodes is 75 and 350 for $T(\Sigma_3)$ and $T(\Sigma_4)$, respectively, with each node occupying 16–24 bytes in memory. For 5 running instances of $T(\Sigma_3)$ and $T(\Sigma_4)$, this amounts to approximately $45KB$ of memory per agent. Additional memory for storing other plan containers such as `OPEN` and `CLOSED` are not considered in this calculation as they store only node references and are cleared after every plan iteration.

Scalability. Our approach scales linearly with increased number of agents. The maximum deliberation time *for all* agents can be chosen based on the desired framerate which is then distributed among agents and their respective planning tasks at each frame. The cost of planning is amortized over several frames and all agents need not plan simultaneously. Once an agent computes an initial plan, it can execute the plan with efficient update operations until it is allocated more deliberation time. If its most imminent plan is invalidated, it is prioritized over other agents and remains stationary till computational resources are available. This ensures that the simulation meets the desired framerate.

9.6.3 SCENARIOS

We demonstrate the benefits of our framework by solving many challenging scenarios (Fig. 9.5)S requiring space-time precision, explicit coordination between interacting agents, and the factoring of dynamic information (obstacles, moving platforms, user-triggered changes, and other agents) at all stages of the decision process. All results shown here were generated at 30 fps or higher, which includes rendering and character animation.

Deadlocks. Multiple oncoming and crossing agents in narrow passageways cooperate with each other with space-time precision to prevent potential deadlocks. Agents observe the presence of dynamic entities at waypoints along their global path and refine their plan if they notice potentially blocked passageways or other high cost situations. Other crowd simulators often deadlock for these scenarios, while a space-time planner by itself does not scale well for many agents.

Choke Points. This scenario shows our approach handling agents arriving at a common meeting point at the same time, producing collision-free straight trajectories. Figure 9.7 compares the trajectories produced using our method with an off-the-shelf navigation and predictive collision avoidance algorithm in the Unity game engine. Our framework produces considerably smoother trajectories and minimizes deviation by using subtle speed variations to avoid collisions in space-time.

Unpredictable Environment Change. Our method efficiently repairs solutions in the presence of unpredictable world events, such as the user-placement of obstacles or other agents, which may invalidate current paths.

Road Crossing. The road crossing scenario demonstrates 40 agents using space-time planning to avoid fast-moving vehicles and other crossing agents.

Lane Selection for Bi-directional Traffic. This scenario requires agents to make a navigation decision in choosing one of four lanes created by the dividers. Agents distribute themselves among the lanes, while bi-directional traffic chooses different lanes to avoid deadlocks. This scenario requires non-deterministic dynamic information (other agents) to be accounted for while making global navigation decisions. This is different from emergent lane formation in crowd approaches, which bottlenecks at the lanes and cause deadlocks without a more robust navigation technique.

Four-way Crossing We simulate 100 oncoming and crossing agents in a four-way crossing. The initial global plans in Σ_1 take the minimum distance path through the center of the crossing. However, Σ_2 predicts a space-time collision between groups at the center and performs plan

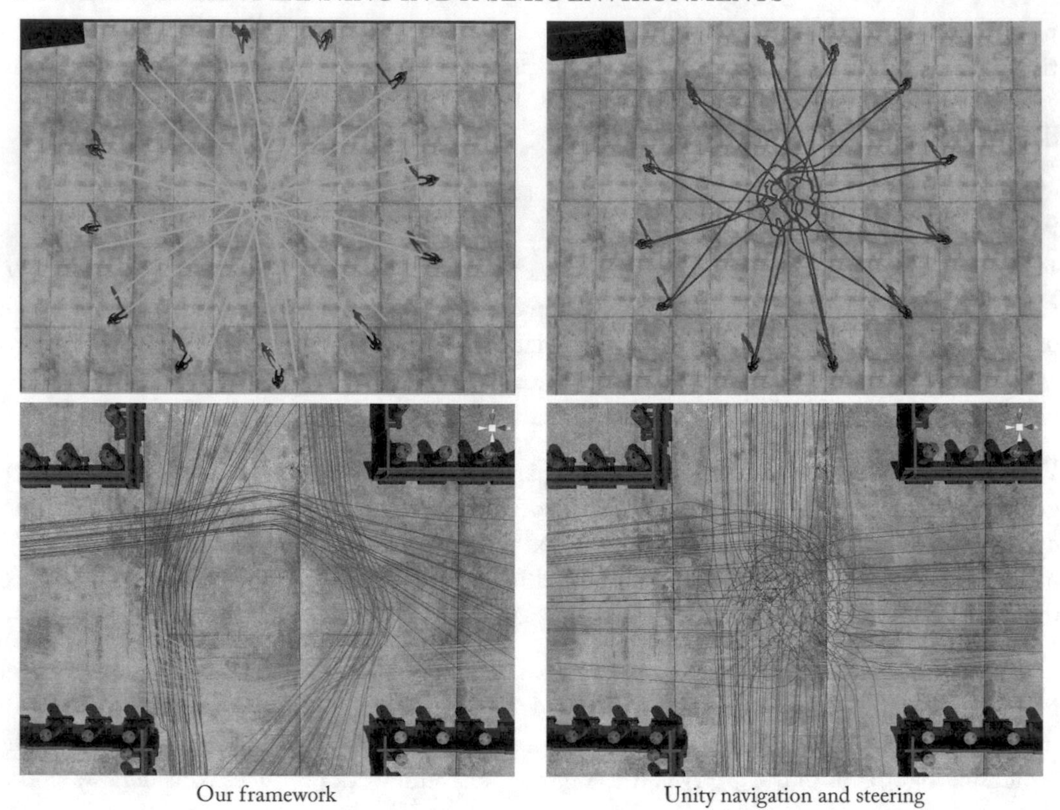

Figure 9.7: Trajectory comparison of our method with an off-the-shelf predictive steering algorithm in the Unity game engine. Our framework minimizes deviation and uses speed variations to avoid collisions in space-time.

refinement so that agents deviate from their optimal trajectories to minimize group interactions. A predictive steering algorithm only accounts for imminent neighboring threats and is unable to avoid mingling with the other groups (second row of Fig. 9.7).

Space-Time Goals. We demonstrate a complex scenario where four agents in focus (additional agents are also simulated) have a temporal goal constraint, defined as an interval $(40 + / - 1 second)$. Agents exhibit space-time precision while jumping across moving planes to reach their target and the temporal goal significantly impacts the decision making at all levels. The space-time domain alone may be unable to meet the temporal constraint and require plans to be modified in earlier domains. No other approach can solve this with real-time constraints.

Many of these scenarios *cannot* be solved by the current state of the art in multi-agent motion planning, which is able to either handle a single agent with great precision, or simulate many simple agents that exhibit only reactive collision avoidance.

CHAPTER 10
Conclusion

We have covered two key aspects of navigation planning for autonomous agents: the importance of having well-defined abstract representations of the environment and planning in multiple domains to achieve paths of high quality with good performance.

We described a fully automatic system NEOGEN that computes a near optimal convex decomposition from a 3D, multi-layered, complex, polygon mesh environment. NEOGEN first does a coarse voxelization of the scene to obtain an approximation of the walkable area. This potentially walkable area is subdivided into layers, using an ordered flooding process, and the layers that are not connected with the user seed, s_w, are discarded. Then, each layer is refined by using the fragment shader at higher resolution and the NavMesh is computed. Finally, all the individual NavMeshes are merged into a single one that represents the walkable space of the entire scene.

The convexity relaxation method is a powerful tool to further reduce the final number of cells, especially when the scenario contains many rounded objects, and hence, the resulting CPG fits well with the requirements of an application that requires real-time performance.

There are other areas for improvement in automatically generating navigation meshes, for example, an ability to handle dynamic events in real time. In applications such as video games, the environment may be constantly changing due, e.g., to an explosion that creates a crack in the floor, a tree that falls and blocks a path, or a door that blocks or makes accessible a region of the scene. In such situations the NavMesh needs to be modified on the fly and paths need to be refined accordingly.

In addition to walking, a character can do more sophisticated actions such as jumping, crouching, or climbing walls. Such abilities allow the character to gain access to parts of the scene that cannot be reached by simply walking, and this information should bc represented on the NavMesh. Automatically labeling portals that require specific actions to be traversed would be a possible approach.

Once a navigation mesh is determined, we can plan trajectories for the agents. The approach described here enables planning in *multiple domains* simultaneously, by leveraging solutions across domains to accelerate computation while still providing a high degree of control fidelity. These domains provide a nice balance between global navigation and space-time planning. Additional domains can be easily integrated (e.g., a footstep domain) to meet application-specific needs, or solve more challenging motion planning problems.

Domains can be connected by using the plan from one domain as a tunnel for the other, or by using successive waypoints along the plan as start and goal pairs for multiple planning tasks. We evaluated both domain relationships based on computational efficiency and coverage, as shown in Table 9.1. Using waypoints from the navigation mesh domain as start and goal pairs for planning tasks in the grid and space-time domain keeps the search depth for Σ_3 and Σ_4 within reasonable bounds. The trade-off is that a space-time plan is never generated at a global level from an agent's start position to its target, thus sacrificing completeness guarantees. This design choice worked well for our experiments where the reduction in success rate of our framework when using this scheme was within reasonable bounds. In balance, it provided a considerable performance boost making it suitable for practical real-time applications. Other applications may opt for different domain relationships.

PART III

Perception

CHAPTER 11

Background

The agent control methods discussed so far have a common feature: movement directives come from essentially omniscient navigation algorithms. Footstep steering uses a local context model based on explicit knowledge of surrounding obstacles and agents, while navigation proceeds with search algorithms exploiting multiple-level spatial abstractions. What real people do when they walk, whether singly or in crowds, involves perceptual processing that triggers responsive and reactive steering and navigation decisions. Crucial components of this "online" process are perception and attention. Within perception, both visual and auditory senses play essential roles.

Furthermore, as crowd simulations become a more accurate depiction of purposeful human populations a more rounded perceptual understanding of the environment is warranted. Physical humans use all of their available senses to interpret the world around them and make behavioral choices based on those interpretations. Does the slight smell of smoke in the house mean dinner is cooking or the house is on fire? What if the smell of smoke is combined with a slight haze rolling out of the kitchen? Those percepts combined with a sounding alarm would have a major impact on the interpretation of the event and reactions to it.

Real human populations interact with the environments they inhabit: They sit on chairs, they eat food, they use computers, etc. Virtual humans should be able to have the same types of meaningful interactions with their environments. To do so, once they are able to perceive their environment, they need to have at least a basic understanding of the objects in the world and how they can be used.

Perception for virtual agents is a diverse topic, as it is a vital part of the sense-think-act cycle found in both physical and virtual modern agents. A large sector of the computer vision community has focused on agent perception, and a survey can be found in [214]. Many of these systems create false color images or saliency maps for agent visual input, the latter being especially used in the computer vision community [30]. Within the virtual agents milieu, vision has been incorporated through top-down perception [293], bottom-up perception [90], and a combination of the two [141]. Visual perception is used for agents steering in crowds, e.g., [200].

Agents with an auditory sense are studied by [90] and [141]. They forgo much of the processing that is required by the vision community, as all the information available is stored directly in the object. Herrero et al. [89] model sound perception in virtual agents by considering sound localization, the sound pressure level of the human voice, and the clarity of the perceived signal. However, the effects of propagation are not modeled and the understanding of speech signals is based on fixed thresholds. Kim et al. [141] models perceptual attention from the top-down and bottom-up with a goal of restricting the sensory information being processed to a realistic level manageable by decision-making components. Sound synthesis and propagation has been of increasing interest in graphics simulations [103, 235, 270]. More important to us is the perception of sound as an influence on agent behavior, e.g., [191] considers audio perception of human voice signals.

Human factors experiments [81] have been conducted to understand the relevance of sound properties for sound similarity: they conclude that amplitude, duration, and frequency strongly correlate with the principal components of sound classification. Based on these human judgments a hierarchical organization of 100 environmental sound signals is generated which clusters sounds that were perceived to be similar. We exploit this model in the sound propagation and perception work which follows in Chapter 12.

Many systems couple visual attention with perception, which allows an observer to determine what the agent is likely to perceive relative to the entire complex of stimuli. The agent could pay attention to events [141] or other agents [232]. Visual attention has also been used to facilitate memory systems [33] and general agent decision-making [293]. While some of these systems provide multiple sensing systems for the agent, they handle each sense separately; that is, there is no attempt to integrate senses. We allow agents to combine different perceptual senses when computing the costs of visual attention, making our system determine sensory, rather than just visual, attention (See Chapter 13).

Perception and visual attention have been added to many different systems for the purpose of creating more intelligent virtual agents. Embodied conversational agents in particular have used perception to allow them to understand the physical or virtual human they are conversing with [213]. This has led to a markup language for perception in conversational agents [237]. Embodied agents can also use perception to interact with a virtual environment, including [90, 141, 293]. Reactive agents, such as those found in [259] also use perception and visual attention to understand events in a virtual environment. [37] and [303] monitor visual attention as a precursor to other motor activities, such as reaching, searching, and catching. [78] include a perceptual workload model for environmental distractions during conversations.

Once agents of a crowd are able to perceive their environment, they need to have enough of an understanding of the objects to make decisions about their use in actions. Such interactions are based on the concept of object affordances, which provides a framework with which to characterize an agent's ability to perceive and interact with its environment. While the notion of an affordance varies in its use within the field of human-computer interaction [184], the definition as originally put forth by psychologist James J. Gibson [71] forms the basis on which this research is motivated. Gibson's affordances possess three defining characteristics [71]:

- Affordances present to an agent potential interactions it may perform with respect to that object's current environment.

- The existence of an affordance is independent of the agent's ability to perceive it.

- The existence of an affordance is binary.

With particular regard for the first point, affordances encapsulate not only the specific actions themselves but also descriptors that may in turn be used to facilitate the performance of action in context.

Smart Objects, so named by Kallmann and Thalmann [49, 116], apply the concept of affordances to objects within a virtual environment. A Smart Object stores its functionality and low-level manipulation descriptors, enabling real-time interactions between the agent and itself. By providing direct and abstracted behavioral and functional representations at the object level [32, 174, 283], this method of affordances provides a base on which to build other high-level descriptions.

Natural language ontologies for the representation of objects and their affordances has been well explored. Bindiganavale et al. [20] pioneered the conceptualization of actions into an ontology to support their execution from natural language instructions. Pellens et al. [211] introduced the use of semantics for the construction of virtual environments at a conceptual level. Kalogerakis et al. [117] proposed the use of ontologies for the structuring of content within objects of a virtual environment using OWL graphs, while Pittarello and De Faveri [221] similarly explored the use of X3D for annotating interactive environments with textual information in a hierarchal form. Balint and Allbeck [7] likewise studied the construction of hierarchies of ontologies using semantic information to facilitate agent behavior. The work in this chapter builds a single framework on a base of existing ontologies for semantically enriched virtual environments.

CHAPTER 12

Sound Propagation and Perception for Autonomous Agents

Pengfei Huang, Mubbasir Kapadia, and Norman I. Badler

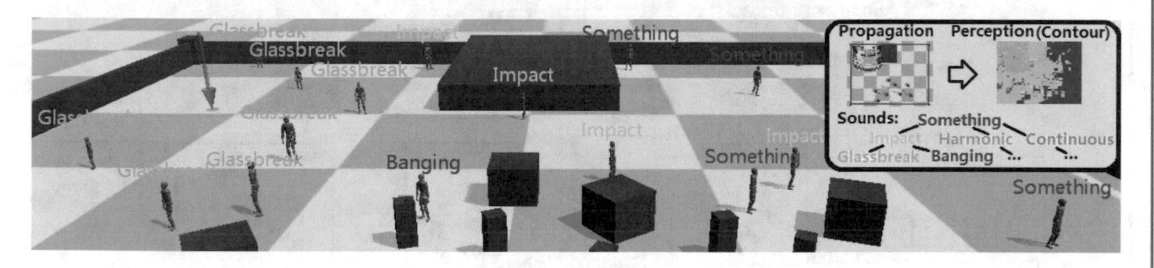

Figure 12.1: Agent-based sound perception using packet representation and propagation model. The green arrow in the scene is the sound source position, and the agents' captions on top show what they just heard. Green indicates correct perception, blue and cyan indicate approximate perception, and red is an incorrectly perceived signal. These sound candidates or categories are from a sound cluster structure.

After looking into multiple perceptual senses for virtual humans, we now turn our attention specifically to audition. Vision-based agent perception uses geometric perception queries such as line-of-sight ray casts and view cone intersections with the environment. However, if we want our virtual agents to behave even more human-like they ought to have hearing models to perceive and understand the acoustic world. Unlike the semantically tagged models of the previous section, we now explore the *passive* sensing of signals from the environment. Moreover, sound propagates differently from light, providing a rich set of additional perceptual options for an agent, including perception and possible localization of an unseen event, and the recognition or possible mis-identification of the sound type. For example, a person may not be seen because of visual occlusion, but the person's footsteps may still be heard. Congested environments and the presence of other agents may muffle sounds to alter or prevent their proper perception.

For virtual reality and games with autonomous agents, acoustic perception can provide useful behaviors including possible goals (sound sources), avoidance regions (noisy areas), knowledge of unseen events (shots), or even navigation cues (such as hearing someone approaching around a blind corner). Virtual agents with "ears" can greatly improve the realism of crowd models, games, and virtual reality systems.

Our approach to sound modeling, propagation, and perception, called SPREAD, is illustrated in Fig. 12.2. First, we describe a minimal yet sufficient set of acoustic features to characterize the human-salient components of a sound signal [81]. These features include amplitude, frequency, and duration, which are found to be strongly correlated to sound classification. Second, we develop a real-time sound packet propagation and distortion model using adaptive 2D quad-tree meshes with pre-computed propagation values suitable for dynamic virtual environments.

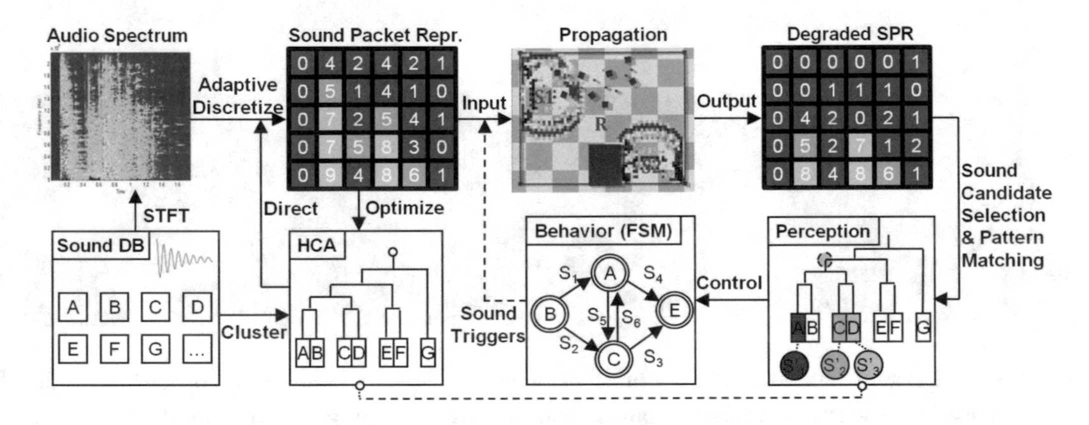

Figure 12.2: SPREAD Framework Overview. (a) Sound signals in a database are adaptively discretized in the frequency and time domain to pre-compute a minimal and sufficient sound packet representation for each sound, and hierarchically clustered based on a human perceived similarity measure. (b) Sounds are propagated in dynamic environments using the Transmission Line Matrix Method to simulate natural sound degradation. (c) Degraded sound packets are perceived using hierarchical clustering analysis to model approximate human-like perception using sound categories. (d) Auditory triggers are incorporated into agent architectures to enhance the behavioral realism of autonomous virtual humans.

During an offline process, we build a sound database using a discrete sound packet representation, and group similar sounds using Hierarchical Clustering Analysis (HCA) based on human sound perception. During simulation, sound packets are propagated through the scene based on the Transmission Line Matrix (TLM) method [108, 145], which accounts for sound packet degradation based on distance traveled, and absorption and reflection by obstacles and moving agents in the environment. To reduce computational costs, we add a quad-tree-based pre-computation to accelerate the propagation model. The algorithm can be easily ported to the GPU for further acceleration.

Agents receive a series of altered sound packets arriving at their location, and use Dynamic Time Warping to identify similar sounds, if possible, from the HCA. If multiple sounds from the HCA are above a similarity threshold, their Lowest Common Ancestor (i.e., the more general sound category) is perceived. Using this framework, virtual agents possess individual hearing. The principle features of SPREAD are:

- An adaptive discretization of continuous sound signals to obtain a minimal, yet sufficient sound packet representation (SPR) necessary for human-like perception, and a hierarchical clustering scheme to facilitate approximate perception.
- Efficient planar sound propagation of discretized sound signals which exhibits essential acoustic properties such as attenuation, reflection, refraction, and diffraction, as well as multiple commingled sound signals.
- Agent-based sound perceptions using hierarchical clustering analysis that accommodates natural sound degradation due to audio distortion and facilitates approximate human-like perception.

Experimental results show that our propagation framework works efficiently for multiple and different sound signals in dynamic virtual environments. A sound signal is easily identified if the agent is close to the source or a sound is less attenuated, absorbed, or reflected in the scene; conversely a sound is difficult to identify as sound packets suffer from interval degradation and overlapping effects. Our sound propagation methodology is not just based on distance, but takes into account the static environment, dynamic features (e.g., other agents), and packet content degradation. We integrate SPREAD into individuals and crowds of agents, demonstrating attention and behavior models in some novel game-like simulations that greatly enhance both play and user experiences.

SPREAD is not intended for auralization (synthesized sound generation), but serves as a virtual companion to auralization—enabling virtual agents to hear and classify sounds much like their real human counterparts. Auralization for humans is not required for this capability.

12.1 SOUND CATEGORIZATION AND REPRESENTATION

Sounds are continuous signals that are typically represented as 1D wave forms. A discretized sound representation must sufficiently capture the distinguishing properties of different signals and facilitate efficient sound propagation in complex environments while exhibiting appropriate sound degradation. This sound data representation will be received by agents who apply human-like sound perception models that determine whether any identifiable sound or sounds have been heard and, if so, what sound type or category they appear to represent. In signal processing, a large number of features are used to represent sound for signal analysis [302]. Human perception, however, is usually correlated to a small subset of features for environmental sounds [81], such as frequency, amplitude, and duration.

12.1.1 SOUND FEATURE SELECTION AND CATEGORIZATION

Sounds attenuate and degrade due to the environmental influences of surfaces that create reverberation, reflection, and diffraction. These effects cause sound signals to degrade in a non-linear fashion, resulting in a lack of perceptual specificity, possible incorrect classification of sound type, or (eventually) complete attenuation. For example, a ship noise may be perceived as a generic mechanical noise, possibly mis-identified as a construction noise which is perceptually similar, but would never be misinterpreted as a harmonic sound such as a siren.

Gygi et al. [81] investigated human categorization of 100 common environmental sounds with an average of 1 second duration, providing a representative sound database. The subjects were required to rate the similarity between any two of these sounds, 10,000 pairs in total, as indicated in Fig. 12.3. Note that there are three clusters with close intra-cluster similarity, and they are later tagged as harmonic sounds, impulsive and impact sounds, and continuous sounds. Based on the similarity matrix, the HCA technique is applied and used to construct a hierarchical clustering of these sounds, as shown in the same figure.

A subset of the full HCA tree is depicted in Fig. 12.3 (c). Perceptually similar sounds are closer in the tree, e.g., *typewriter* and *keyboard* sounds are under the same node and their HCA distance (tree-edge) is two units (one unit from *typewriter* to its parent node plus another one from the parent node to *keyboard*). The distance metric applies to any two sounds in the tree. The branch nodes are named to describe the meanings of a cluster of sounds that are under that particular branch; e.g., *gun* and *axe* sounds are clustered as *destructive* sounds. All right-side sounds are *single impact* sounds, and the overall tag is *impulsive* for all the sounds in this figure.

These 100 sounds provide a carefully chosen, representative set of common environmental sounds, and we leverage existing perception studies [81] to ground our approach in actual human perception. The sound duration is limited to about 1 second which is long enough for a distinct sound event; sounds with longer durations can be segmented and processed in sequence. SPREAD can be easily extended to new sounds by importing raw sound data and extending the HCA tree. The clustering information can be acquired from existing studies, running new human subject experiments, or manual labeling.

12.1.2 SOUND PACKET REPRESENTATION (SPR)

A sound signal is traditionally represented by a wave-time or spectrum-time graph which models three fundamental features: amplitude, frequency, and duration. SPREAD employs a packet-based discretization of sound: the Sound Packet Representation (SPR) based on the Short-Time Fourier Transform (STFT) [98]. SPR packets can be efficiently propagated in discrete grids.

In Fig. 12.4, we show that we can reduce the number of packets by using fewer frequency bands and only storing packets for sound segments with a significant amplitude. Along the hor-

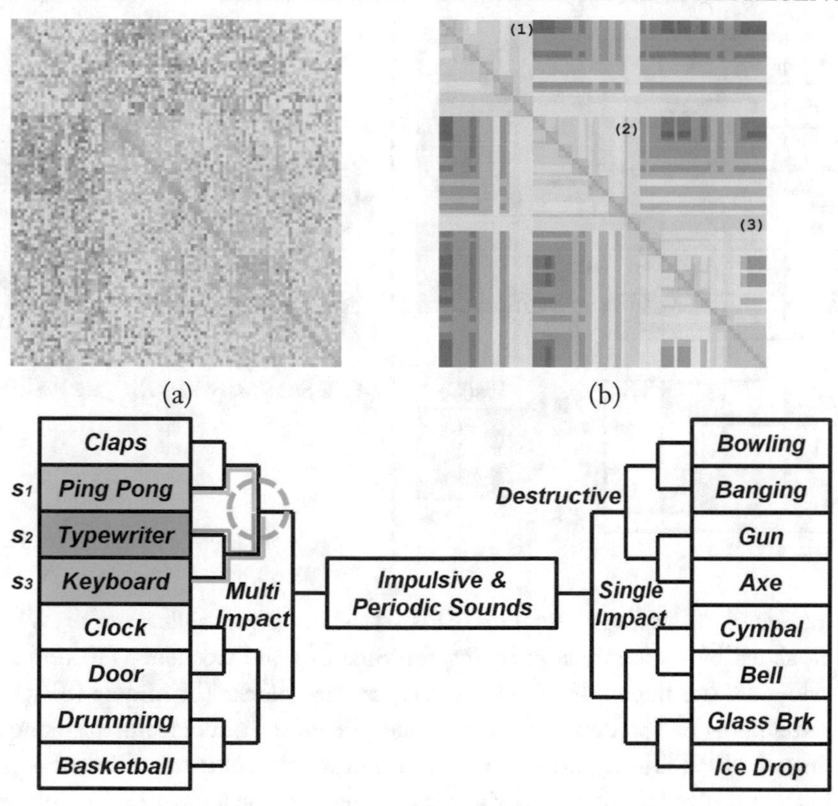

Figure 12.3: Sound Perception Similarity Matrix: (a) is the sound similarity matrix constructed from human evaluations, while (b) is calculated according to the HCA tree structure. A green comparison block means that people think two perceived sounds are similar as in (a); green in (b) means the sound nodes are closer in the tree; (c) shows that the similarity matrix (a) and its clustered block counterparts marked in (b) can be transformed into a partial hiearchical clustering tree (HCA) via Multidimensional Scaling [25]. Note that if multiple sounds are identified as similar to a given signal, we hypothesize that people will perceive that signal as a coarser category which is the least common ancestor (the yellow circle) of the sounds ($s1$, $s2$, and $s3$). This idea will be described in detail in the perception section which follows

izontal axis, we represent a signal as a time-varying packet sequence. Either one packet with one amplitude value in a frequency range is generated at a time step or multiple packets for various frequencies are generated at each time step.

In SPREAD, a packet $p(i, j)$ is denoted as $\langle a, \mathbf{r} = (r_L, r_H), s \rangle$, where i is the time axis index, j is the band index along the frequency axis, a is the amplitude, r_L is the lower bound of the perceptive frequency band, r_H is the upper bound, and s the spread factor which defines the

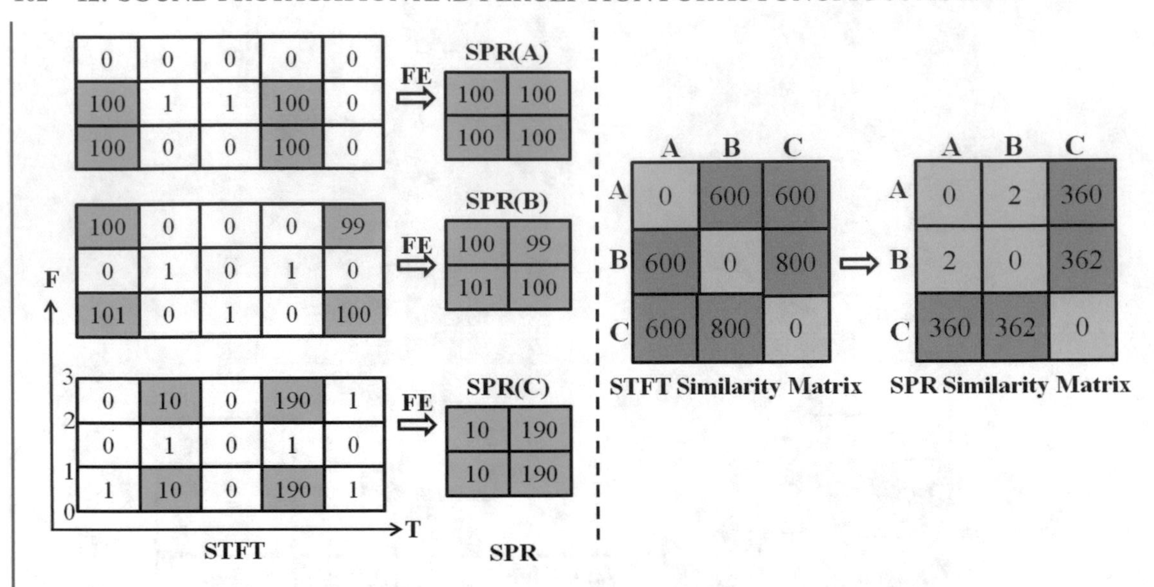

Figure 12.4: Sound Packet Representation (SPR). The left diagrams illustrate the STFT conversion of the sound signal by uniformly segmenting the time (T) and frequency (F) domains, where the numerical values are the amplitudes within each discretized block. The middle (SPR) column shows the feature extraction (FE) process which determines the most distinct features among all blocks that are packed into the SPR. The right diagrams show that we can construct different representations in order to find an optimized similarity matrix that best matches the known HCA clusters.

degradation extent of the packet. Thus, a sound signal is represented as a collection of packets $\{p(i,j) \mid t_0 \leq i \leq t_n, b_0 \leq j \leq b_m\}$. The time duration of the sound signal is the total length of the packet series $t_n - t_0$ along the time axis.

In 12.1.3, we describe the algorithm to extract the minimal yet sufficient set of crucial and necessary packets in M frequency bands and N time slots from the original STFT data to represent a sound and the correct clustering of sound categories. Overall SPREAD efficiency is improved if unnecessary packets (relative to the selected sound database) are eliminated prior to propagation.

12.1.3 SPR SELECTION FOR HIERARCHICAL CLUSTER ANALYSIS

SPR is for a single sound whereas HCA is for multiple sounds. To establish a meaningful relation between these two we must define a comparison measure within the simulation. What we want is that sounds under the same or close clusters can be measured and evaluated as perceptually similar, while those which are more distant should be judged as different. The chosen comparison measure must operate on any subset of the sound packet data, since packets from the same or multiple sources may overlap at the receiving agent's location.

Dynamic Time Warping [282] is a technique to compare two time sequences, and SPREAD uses it to determine the similarity between two sounds in wave or STFT spectrum forms. Note that packet sequences will all have $\|N\|$ frames and within each frame there will be M packets in different frequency bands. The difference between any two frames are defined as the sum of all the corresponding packet pairs (i.e., $d(f_j, f_k) = \sum q_{mi} - r_{mi}$). The distance $DTW(s_a, s_b)$ between two sound signals s_a, s_b is computed by applying the Dynamic Time Warping algorithm to the packet sequences in s_a and s_b, as in Eq. 12.1:

$$DTW(s_a, s_b) = min\{C_p(s_a, s_b), p \in P^{len(s_a)\times len(s_b)}\} \tag{12.1}$$

where $len(s_a), len(s_b)$ are the total number of time frames of s_a, s_b respectively, $P^{len(s_a)\times len(s_b)}$ is the set of all possible warping paths in the cost matrix $d(i,j)$ and $C_p(s_a, s_b)$ is the cost of two sequences along the path p which is the min-cost frame-to-frame mapping between them from the beginning to end along the time indices. The amplitude a, frequency band range $\mathbf{r}$, and spread factor s are used to compute the metric difference between any two packets (i and j) in the sequences: $d(i,j) = \mu(a_i - a_j) + \nu(1 - \frac{\mathbf{r}_i \cap \mathbf{r}_j}{\mathbf{r}_i \cup \mathbf{r}_j}) + \xi(\frac{\|s_i - s_j\|}{\|s_i + s_j\|})$. In our problem setting, we have $\mu = 100$, $\nu = 1$, and $\xi = 1$.

Our requirements for SPR are that it:

1. be minimal yet sufficient (it needs to be the minimum representation that can sufficiently distinguish between all leaf nodes in the HCA tree);

2. should not be so fine that it wrongly discriminates similar sounds in the same category;

3. should be computationally efficient (so a smaller subset of data may be used).

Thus, we seek to find optimal representational subsets of the data, as shown in Eq. 12.1 and Eq. 12.2. In Eq. 12.2, t denotes a tree node, R_t is the regularization value on t, R is the total, (a,b) denotes all sub-leaf node pairs under the tree node t, D is the DTW function as defined above.

$$\begin{aligned} R &= \sum_t R_t \\ R_t &= \begin{cases} 0, & \text{if } t \text{ is a leaf node} \\ \frac{\sum_{(a,b)} DTW(t_a, t_b)}{\#of(a,b)pairs}, & \text{if } t \text{ is not a leaf node} \end{cases} \end{aligned} \tag{12.2}$$

Our SPR framework selects representative frequency and amplitude features from audio clips, but the sparse sampling may fail to capture salient differences that a person would normally perceive, while dense sampling may introduce noise and error and also fail to show distinct differences. To minimize this ambiguity we use an algorithmic feature selection process based on human sound perception. Feature selection is optimized to match the target sound perception space stored in the HCA tree structure, so that the difference between any two signals has a distance similar in scale to the corresponding two nodes of the HCA tree.

Figure 12.5 (a) shows that the ordering of sound signals based on HCA tree distance can differ from the ordering based on the distance computed using DTW, resulting in incorrect perceptual clusters of sound signals. To offset this issue, we must select features of the sound signal by sampling at specific time slots such that the computed distance aligns with the perceived difference. We choose a set of sampling slots N, M along the time and frequency axes such that the DTW distance between all sound signals is aligned to their HCA distance.

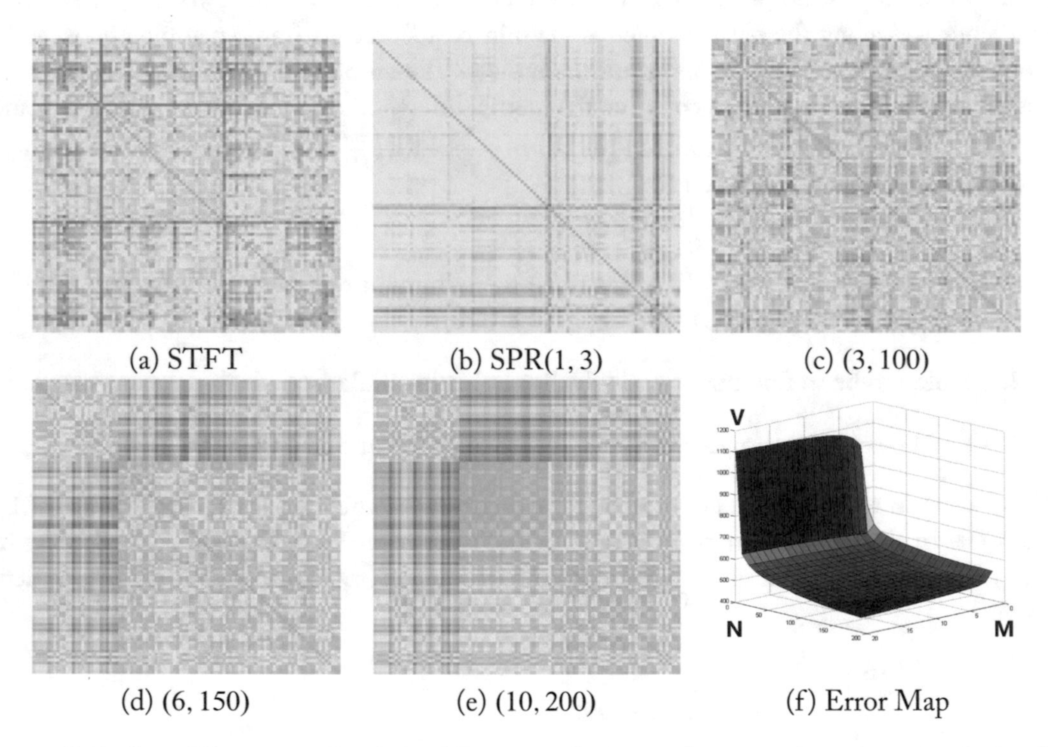

Figure 12.5: Sound Perception Similarity Matrix, in which sound-to-sound similarity is calculated by DTW on the following different datasets (M is number of frequency bands and N is number of time samples): (a) the original STFT data consists of, on average, 500 time slots and 512 frequency bins; (b) SPR data constructed from HCA using M = 1 and N = 3 making the sounds hardly distinguishable; (c) using M = 3 and N = 100; (d) using M = 6 and N = 150; (e) using M = 10 and N = 200. In (f), the curved surface image shows that if M and N are set to their maximums (10 and 200, respectively) then the similarity error V is minimized. Lower M and/or N increase error. A suitable error tolerance can be set at the application's discretion.

If we compute the similarities among the complete dataset (shown in Fig. 12.5 (a), the result differs too much from human perception and will give unsatisfactory or implausible matches. After the optimization and construction algorithm, the similarity is shown in Fig. 12.5 (e) which matches well the human subjective results. The factor analysis is shown in Fig. 12.5 (f).

12.2 SOUND PACKET PROPAGATION

Many sound propagation methods exist, such as FDTD and FEM in the Numerical Acoutics (NA) field, and ray-tracing and beam-tracing in the Geometry Acoutics (GA) domain. Here we use a rectilinear cellular space that approximates the physical environment of static and dynamic objects and agents, and propagate sound by the TLM Cell-automata Acoustics (CA) model. CA's computational cost is independent of the number of agents; GA increases per agent. Also, GA physically approximates sound waves as lights, making the GA diffraction model expensive, whereas CA inherently models all sound and environment interaction effects.

12.2.1 TRANSMISSION LINE MATRIX USING UNIFORM GRIDS

The sound signals received by agents depend on the cell they occupy, but their other actions (such as navigation) are not restricted to this grid. Changes in a packet's feature value are governed by known formulas for sound propagation effects. For a detailed review of the TLM algorithm on uniform grids see [145].

The TLM method belongs to the Cell-automata Acoustics (CA) category along with other methods such as Lattice-Gas and Lattice-Boltzmann models. TLM is based on Huygen's principle, as shown in Fig. 12.6 (a) & (b), where each point in the wavefront is a new source of waves. Given a grid-based discretization of the ground plane of the scene, the sound distribution can be calculated by first updating the current energy values for each grid cell. Then for each neighbor of each grid cell the energy is calculated that will be transferred from the center grid to the neighbor grid, as shown in Equation 12.3. To model reflection, the vector values which reach a wall will simply reverse direction.

TLM can simulate multiple simultaneous sound sources anywhere in the grid. Packets at the same location and the same bands will be merged and their amplitudes will be added. The TLM grid can also represent "constant" ambient sounds (e.g., general levels of traffic noise) that just sum with any transient packets. Such levels can be ascertained empirically or from other simulations [202] and create appropriate perceptual confusions. Moreover, the agents themselves can generate local sounds (*footsteps*, *handclaps*, or *non-linguistic utterances*) in the grid. Agent presence increases the grid absorption but does not otherwise impact the propagation algorithm.

$$
\begin{aligned}
a(g_i) &= \alpha(g_i) \cdot \left(-\frac{a(f_{op(i)})}{2} + \sum_{j \neq op(i)} \frac{a(f_j)}{2}\right) \\
a(Neighbor(g_i)) &= a(g_i)
\end{aligned}
\tag{12.3}
$$

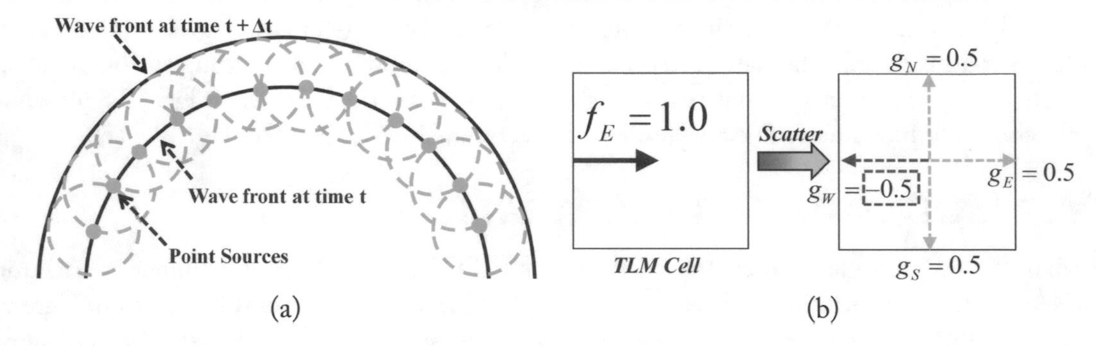

Figure 12.6: (a) Illustration of Huygen's principle: starting from a spherical point source, the wavefront in the next time step is formed by propagation at the border of the current one. (b) Grid-based sound packet propagation: An original incoming packet (the energy with an arrow pointing to the center of the TLM cell) will scatter into four subpackets ($f_E \rightarrow \{g_N, g_E, g_S, g_W\}$) which are the outgoing packets to be transmitted to its neighboring four-connected cells (N,E,S,W), where they become new incoming packets in the next time step.

$$
\begin{aligned}
s(g_i) &= \sum_{j \neq i} s(f_j) \cdot \delta_s(g_i) \\
\delta_s(g_i) &= \begin{cases} 0.01, & \text{if } g_i \text{ is an agent grid} \\ 0.10, & \text{if } g_i \text{ is a wall grid} \\ 0.98, & \text{if } g_i \text{ is an ordinary grid} \end{cases}
\end{aligned} \tag{12.4}
$$

$$
s(f_i') = \texttt{Collect}_s(i, \{s(g_0), s(g_1), ..., s(g_k), ...\}) \tag{12.5}
$$

We extend the scatter rule in Equation 12.4 to work for the sound packet's spread factor. Here, $s(g_i)$ is the spread factor which indicates how clear or fresh the packet is at grid g and direction i, and $\delta(s)$ is the decrement multiplier for the factor. The collection rules in Equation 12.5 merges the incoming packets together with their spread factors merged (summed). Note that SPREAD does not fundamentally change TLM, but we propagate only key packets, model their interactions, and track their spread (degradation) during propagation.

12.2.2 PRE-COMPUTATION FOR TLM USING A QUAD-TREE

For a square region with a uniform sound attenuation property, the propagation pattern is always the same and is proportional to the original source energy, which can be pre-computed and cached as shown in Fig. 12.7. Moreover, since the sound signals we are processing are fairly short

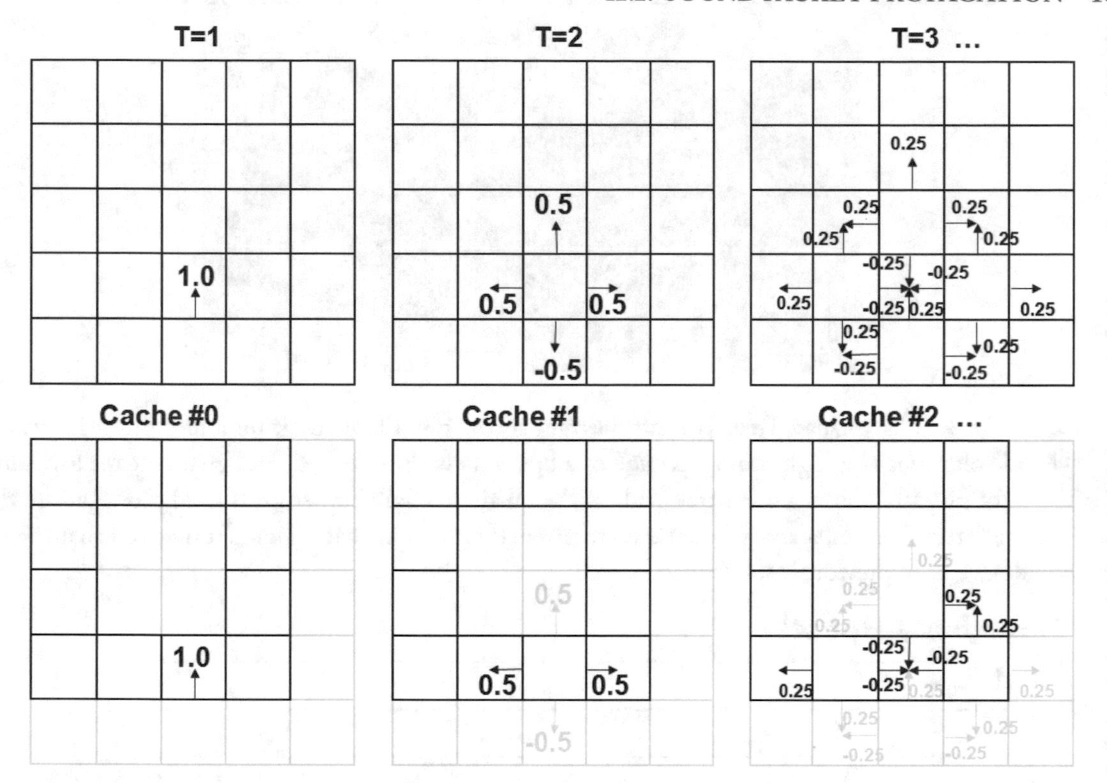

Figure 12.7: Using the TLM pre-computed cache. The top row shows a number of frames for propagation in the full domain. The bottom row shows a highlight view of only the border grids of a quad region (here 4x4). For a unit incoming trigger packet, its consequent propagation pattern is deterministic so can be pre-computed and cached. For any incoming packet with value v, multiply v by the cached values and apply them to future frames.

with only about 150 frames of packets, any sound trigger only needs to be propagated for less than 3 seconds. This particular constraint allows us to cache some of the propagation results and accelerate the overall algorithm.

We subdivide the entire scene into quads such that each quad region has uniform acoustic properties. Given any input sound, we can find the relevant propagation pattern and its result, and then assign the distribution values to the incident grids, repeating this process for each timestep. The propagation results using this method are identical to using a uniform grid, as illustrated in Fig. 12.8, and provides a tremendous performance boost (Fig. 12.9).

To explain why QTLM gives approximately linear runtime (with regard to the log scale of space resolution as shown in the Figure 12.9), UTLM on $R * R$ grids has complexity of $O(R * R)$ because it needs to update all its grids, but QTLM only needs to update the borders of grids,

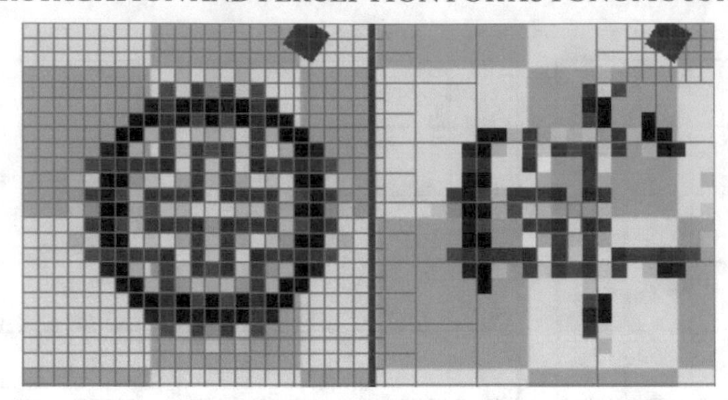

Figure 12.8: TLM Quad Tree. The left diagram shows the TLM result on a uniform grid, where the red color means a high sum amplitude of all packets within the cell, and green means low sum. The right diagram shows a quad-tree grid. In the quad-tree only border grids need propagation: the "internal" grids are unnecessary because no receivers exist within that region. (If they did that region would have been previously subdivided.)

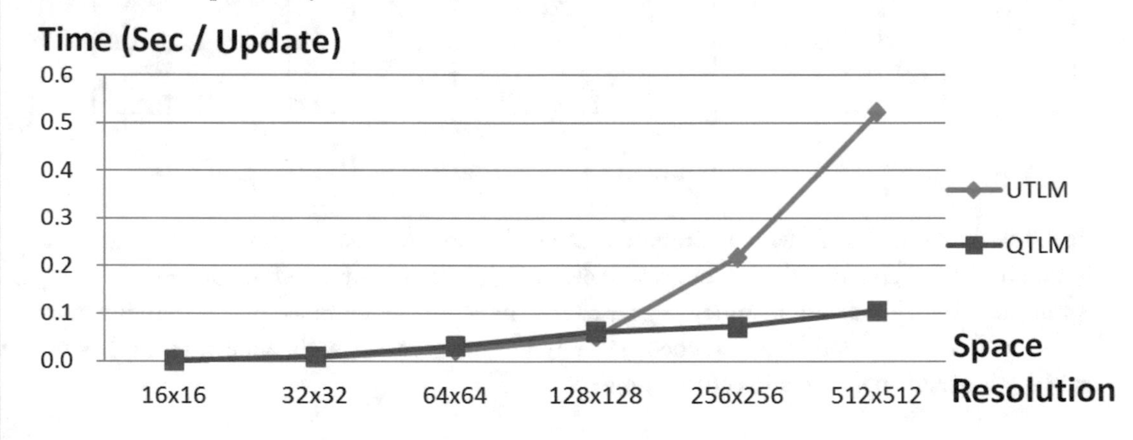

Figure 12.9: Performance comparison between Quad-tree TLM and Uniform TLM. Computational cost of UTLM increases quadratically, while QTLM increases only linearly. For a 512x512 resolution, QTLM takes about 5G memory and 15 seconds overhead pre-computation on an x64 machine. But with further optimization, these constraints could be much reduced.

which is approximately $O(R)$ because only the border grids count. Then for each border grid's effect, it needs to update T (at most $4 * R$ for a quad) other border grids in L consecutive frames, $O(T * L)$ in total. Since practically a single packet will not impact most of the border grids or most of the following consecutive frames within the current frame, $T \ll 4 \cdot R$, much less than $O(R)$ and moreover $L \leq R$, and in total $O(T * L) \ll O(R * R)$. In fact, the larger the value R is, the more runtime will be saved (because $L \ll R$ then) with the trade-off of greater (but one time) overhead of pre-computation time and more cache storage. Updates on quads without packets are obviously unnecessary. In total, the combination gives approximately $O(R * T)$ ($O(R)$) performance, also reflected in the chart. Furthermore, quad-tree pre-computation is only for each different size of obstacle-free quad (1x1, 2x2, 4x4, 8x8,...), and the pre-computation does not need to consider any nested tree configurations. The limitation of this algorithm is that it is not suitable for long duration propagation because L will be very large and the computational and space cost will be very expensive. However, only high amplitude sounds will create large L. Based on existing algorithms [162], quad-tree updates can be achieved in $O(1)$ complexity, as long as the dynamic changes only affect neighbor grids. This fits with our dynamic simulation framework for autonomous agents.

12.3 SOUND PERCEPTION AND BEHAVIORS

Hearing helps us experience, communicate with, and react to the ambient environment and other people. Leaving aside linguistics, we can still build a sound perception model for virtual agents so they can identify, as well as possible, any environmental sound packets they receive.

12.3.1 EFFECT OF SOUND DEGRADATION ON PERCEPTION

Agents perceive two types of information from any packets that arrive at their ground location: the impulse responses of packets at different frequency bands and the spread factor that is computed for each packet which indicates the frequency and amplitude changes due to environment interactions and attenuation. Then we use Dynamic Time Warping (DTW) [282] to compute similarity values between these packets and all the sounds in the HCA database. The similarity value ranks possible leaf node matches or probable general categories related to the spread factor. The process is shown in Fig. 12.10.

12.3.2 HIERARCHICAL SOUND PERCEPTION MODEL

Agents should be able to identify clear (or nearby) sounds accurately, but degradation may confound accurate identification. We exploit this to model sound perception based on the HCA tree structure. For example, *ice drop* and *glass break* are grouped as a single *impact* sound which is *non-harmonic* and *impulsive*. *Blowing*, *gun*, and *axe* sounds are grouped together into *destructive* in their common ancestor node, and other sounds such as *clocks*, *drums*, *claps*, and *typewriter* are grouped as *multiple impact* sounds.

If an agent is unable to accurately perceive a sound (a leaf node) due to packet degradation, it may still find a similar sound type at a coarser level in the HCA tree. An unintelligible sound would map to the root node of the tree: the agent hears something but cannot identify what it is.

An agent may receive a temporal series of packets from more than one source. The packet series of each sound in the database are compared with the received series using DTW. Since there may be multiple sound sources, the first matched packets will be removed from the received

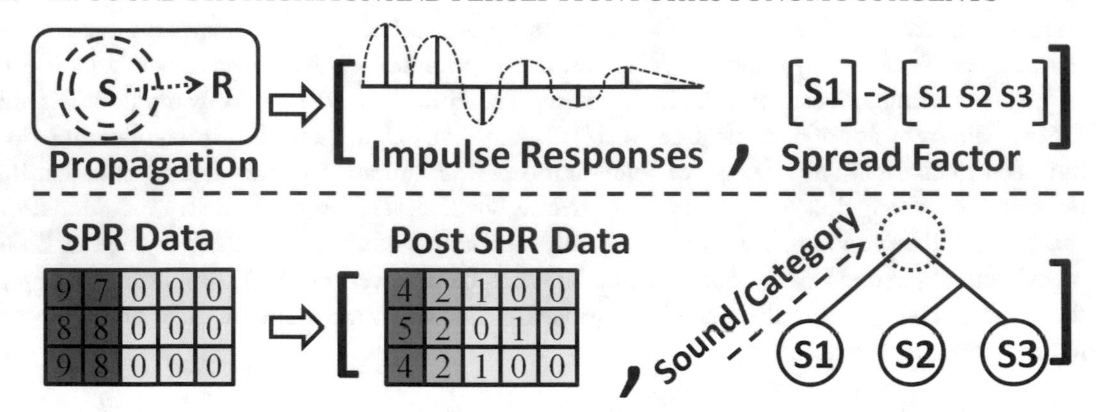

Figure 12.10: Post-propagation data used for sound perception model. The top row shows that after propagation, the impulse responses (IR) of the original sound signal sequence (from SPR) will be received along with the spread factor indicating any frequency degradation. The bottom row shows how post SPR data, which is computed based on IR, resulting from degradation and change during propagation, affects the perception of sound category.

packet set, so the extract and match processes can continue with the remaining ones. For example, in a series of received packets, suppose there are three distinct sets, and two of them are in the high bands and belong to the *siren* sound, then they will be used to match first. The remaining (one) low band will be used to compare with other sounds. Note that this greedy step will introduce error. Although people are good at distinguishing commingled sounds, a perfect blind source separator is difficult to model [11].

The spread factor value models an SPR's range dispersion. The smaller the spread factor, the more degraded and approximate will be the perception. The relation between spread factor and candidate number K is chosen to be linear, though other relations could be used instead. Assume the reference spread factor is S (i.e., 10), then if any sequence's specific spread value sum (of all the received packets) is more than 95% of S, then only the top (first) candidate will be considered and used to find the HCA node; if more than 90% then the top 2 candidates, 85% the top 3, and so on.

In terms of identifying the perceived sound information, the top K (≥ 1) candidates below a similarity comparison threshold (least similarity * 10) are output as the set of perceived sounds. These sounds map to leaf nodes in the HCA tree structure, and we define their least common ancestor as the perceived sound category. As shown in Fig. 12.3 (c), a given sound signal has similar sounds s_1, s_2, and s_3, and so the perceived sound category is their least common ancestor. Figure 12.1 illustrates the perception results of the *glass break* sound at different locations: (1) open area, (2) high absorption region, (3) high reflection region, (4) blind corner, and (5) sound blocking region. The *glass break* sound is clearly heard in nearby or open areas, with coarseness of perception increasing in complex surroundings with obstacles and other agents. In contrast, a harmonic sound like *harp* is accurately perceived in most of the areas. These examples show that SPREAD accounts for sound characteristics and the dynamic configuration of the environment, and is not a simple distance-based perfect reception function.

12.3.3 SOUND ATTENTION AND BEHAVIOR MODEL

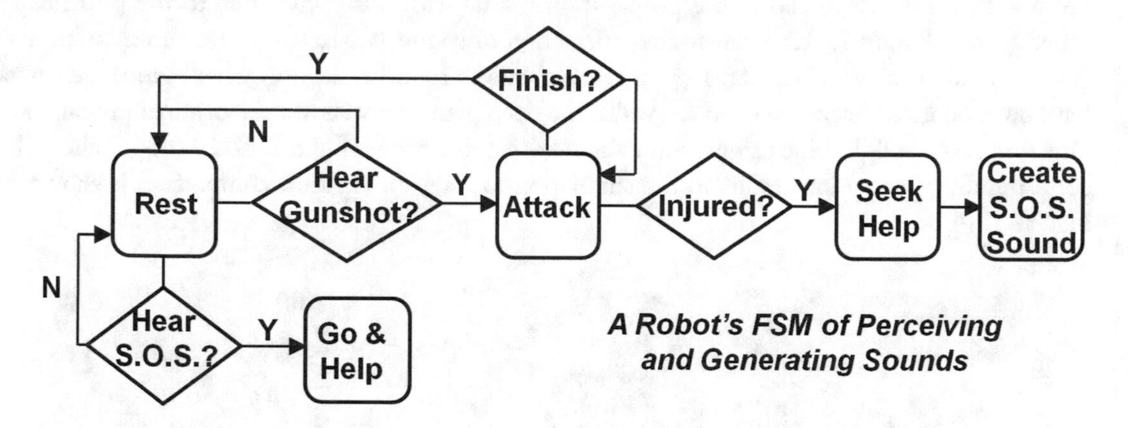

Figure 12.11: Finite State Machine example for modeling an agent's behavioral response to auditory triggers.

An agent's response to sound depends on being able to hear and possibly disambiguate it from noise, but it also crucially depends on a human cognitive property: attention. We use a very simple model based on two attention measures. The first is the amplitude threshold $A = 100$. When any sound's total amplitude, the sum of all packets during a number of frames ($\simeq 150$), exceeds this limit it will definitely draw one's attention. Sounds which have low amplitudes are unintelligible or just contribute to nondescript background noise. The second measure computes the saliency or conspicuity of a new sound by comparing its packets with those of the previous sounds: if the difference between them is greater than a percentage threshold $P = 30\%$, it also triggers the agent's attention. These attention triggers can be used to select, modify, or terminate associated agent behaviors. Figure 12.11 illustrates a simple finite-state controller for agent steering behaviors based on audio perceptions. Other agent control mechanisms are clearly possible and can be embedded in simulations or games.

12.4 EXPERIMENT RESULTS

For our experiments, we use the same set of 100 environmental sound signals that were used in [81]. We demonstrate SPREAD using a simple virtual environment with static obstacles and moving agents. Static obstacles occupy grid cells with absorption and reflection rates for sound propagation. Moving agents dynamically map their absorption and reflection rates to the grid cell they presently occupy. Our system is built on top of the ADAPT platform [246] (Section 21) which provides tools for global navigation, goal-directed collision avoidance, and full-body character animation.

Figure 12.12 illustrates the different acoustic properties implemented in SPREAD. All these results arise from a single omni-directional sound emission pulse at the purple point in the images. Figure 12.12(a) shows the propagation results with absorption due to the presence of other agents. Figure 12.12(b) illustrates diffraction of sound where the green automata propagates around an obstacle corner. Figure 12.12(c) shows sound reflection where green automata have been bounced back from the red walls; the deep blue arrow shows the original propagation direction and the light blue arrows show the reflected directions. Figures 12.12 (d)–(f) show the corresponding perception results in the simulation for (a)–(c). A result comparison is shown in Fig. 12.13.

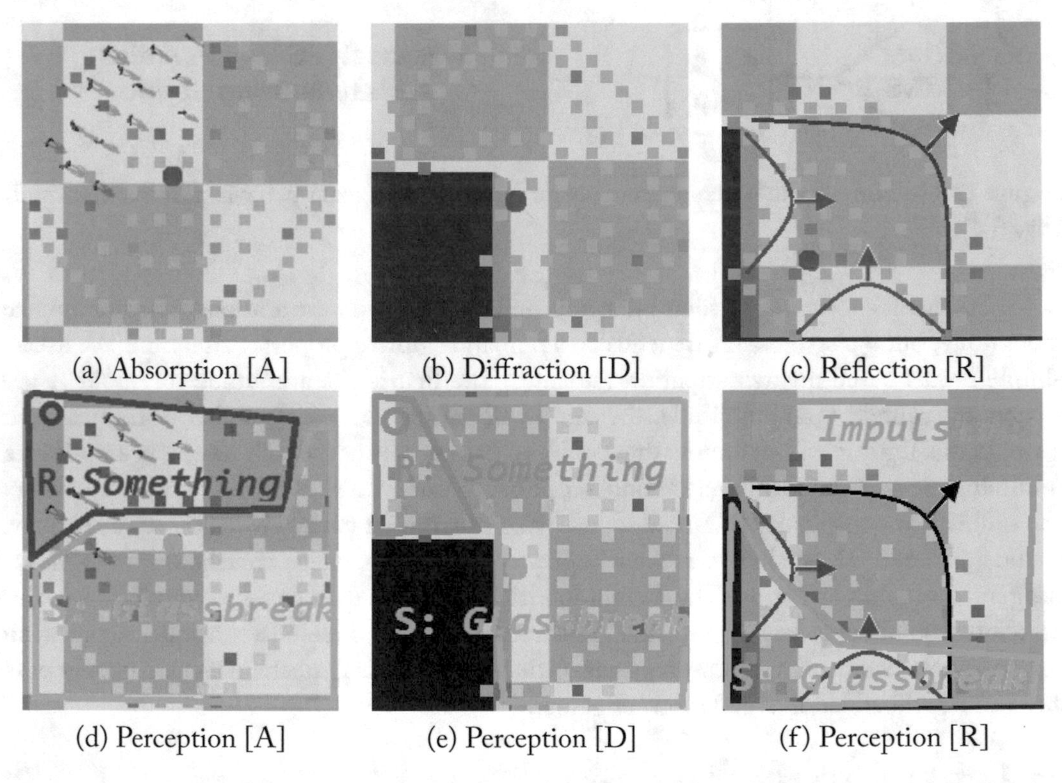

(a) Absorption [A] (b) Diffraction [D] (c) Reflection [R]

(d) Perception [A] (e) Perception [D] (f) Perception [R]

Figure 12.12: TLM propagation results showcasing different acoustic properties.

Figure 12.14 illustrates the perception contour map using SPREAD. We observe a non-linear separation between regions of accurate, approximate, and incorrect perception. This is in

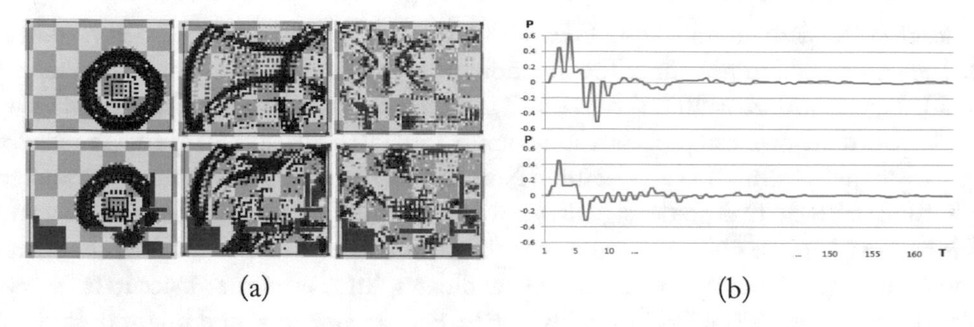

Figure 12.13: This figure shows the similar results from our TLM propagation and Raghuvanshi et al.'s FDTD method on the emission of one Gaussian impulse. (Refer to Figure 6 in [224].) (a) the propagation results on two different scenes of a same impulse packet; (b) the impulse response packet values sampled along time frame at the receiver (near the source). Note that this is only for one frequency band.

sharp contrast to existing models which generally use simple distance-based functions, highlighting the veracity of our approach.

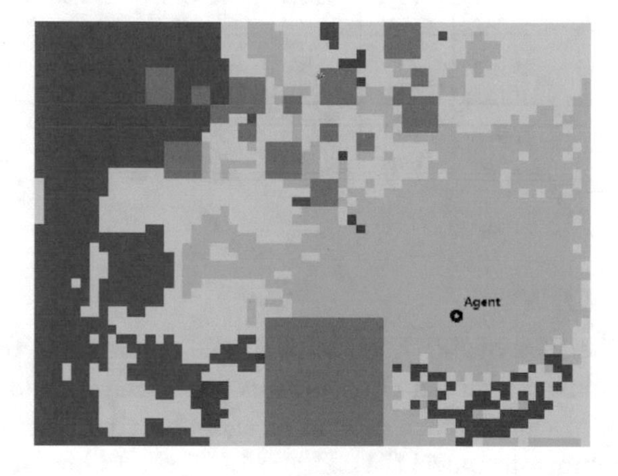

Figure 12.14: Perception contour map for one agent listening to a *glassbreak* sound played at different locations in the environment with reflective obstacles (gray).

The accuracy of sound perception with respect to the sound packet representation parameters are shown in Table 12.1. In the experiment, there are 30 agents, and we play the sounds at various points and capture each agent's perceptual matches. If the perceived sound is exactly the played sound, it will increase P_a which is the percentage of accurate perceptions, and if the perceived sound is an ancestor (in the HCA tree) sound it will increase P_f which is percentage of approximate perception. For $M = 1$, the accuracy is very low, because adding all the spectrum values together destroys the distinct features of each frequency band. The experiments were run on a desktop PC with an Intel i7 2.8GHz CPU, 16GB RAM, and a Quadro NVS 420 graphics card. The system runs in real-time. The situation shown in Fig. 12.1 uses parameter values $M = 3$, $N = 50$, #agents=20, $\mu = 100$, $\nu = 1$, $\xi = 1$, and the effective length of sound data is typically about 1s. Currently, we can process a scene up to 150*150 grids and 50 agents in real-time, and can greatly benefit from GPU acceleration. Note that the "accurate" perception percentage is only 43.5%. Since SPREAD degrades signals, we would actually be more surprised if this were higher. This is because here we only output one (the "best") result, and do not consider the spread range to find a more general category for a set of candidates. Thus even the "inaccurate" ones are still similar to the original's siblings in the HCA tree; e.g., an engine sound might degrade to a mechanical one (typically low frequency and non-periodic), but much less likely to a siren-like sound (high frequency and periodic). By considering these degraded perceptions, our algorithm gives a reasonable percentage of "accurate+approximate" perceptions.

Table 12.1: Sound perception accuracy. Accuracy statistics of sound perception for varying N and M (the number of samples along the time and frequency axes, respectively)

	Acc.	$N = 10$	$N = 50$	$N = 150$
$M = 1$	P_a	0.0%	0.0%	1.0%
	P_f	22.7%	32.4%	28.7%
	P_{total}	22.7%	32.4%	29.7%
$M = 3$	P_a	19.9%	43.5%	38.0%
	P_f	1.0%	46.3%	50.9%
	P_{total}	20.9%	89.8%	88.9%
$M = 6$	P_a	36.1%	40.7%	40.7%
	P_f	32.4%	53.2%	57.8%
	P_{total}	68.5%	93.9%	98.5%

12.4.1 APPLICATIONS

We demonstrate the benefits of SPREAD with simple applications that showcase the role of auditory triggers in interactive virtual environments. The behavior models for the autonomous agents are simple state machines which depend on agent hearing. They can be easily replaced by more complex behavior architectures [131]. Figure 12.15 shows how sound triggers can be used to attract the attention of another agents and affect its approach to a blind corner.

SPREAD has also be applied to an existing game application. The original game without sound perception involves a player-controlled avatar searching for and destroying enemy robots in a maze-like environment. The ability to perceive sounds greatly enriches a simple game mechanic where robots perceive and react to different sound signals in the environment. Robots hear a gunshot and retreat or attack depending on their health status. They can additionally cry out for the assistance of nearby robots, by triggering a sound signal. Players can also mimic the robot cry to lure robots to an isolated location. The resulting gameplay is greatly diversified where players use a stealth mechanic to isolate and corner robots in cordoned off areas where other robots are unable to see and *hear* them.

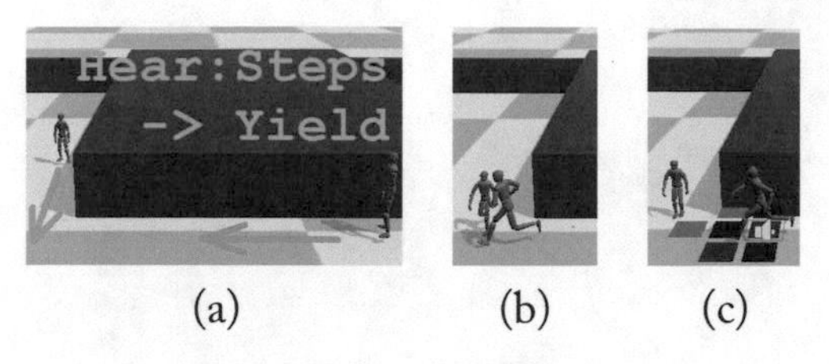

(a) (b) (c)

Figure 12.15: Blind Corner Reaction Example. (a) Agents walking toward each other at a blind corner. (b) Without sound propagation or perception modeled, agents bump into each other. (c) With correct models, one agent can perceive footstep sounds and yield.

SPREAD is currently limited to grid-based discretizations of the environment where the grid resolution has a combinatorial effect on the complexity of the TLM-based algorithm for sound propagation. This limits the applicability of our approach to relatively simple block worlds while meeting real-time constraints. Parallelization and the use of triangulations are possible future directions for increasing the computational efficiency of our method.

CHAPTER 13

Multi-sense Attention for Autonomous Agents

John T. Balint and Jan M. Allbeck

13.1 INTRODUCTION

If the primary role of virtual agents in a crowd is to behave like real humans, they ought to make human-like decisions. As physical humans interact within the bounds and understanding of their environment, which can change unexpectedly, virtual humans should be able to as well. To simulate this, an agent must be able to perceive and understand what is going on about them, as illustrated in Fig. 13.1. As one of our goals is to make more purposeful, functional crowds, agents' responses to and interactions with objects in the environment are critical.

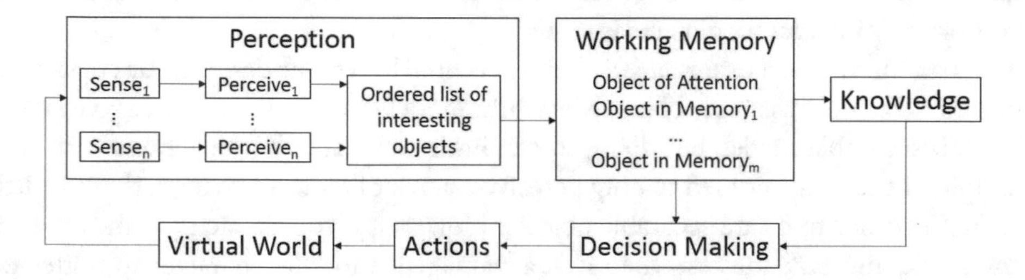

Figure 13.1: A sample agent framework with the perception and working memory portions expanded. Here the perception module is decomposed into a number of sensing abilities.

An agent interacting within a semantics environment, also known as a SmartObject environment [49, 115], must be able to use and understand the meaningful information found within it. Complex computer vision systems have been designed to imbue an agent with synthetic vision, allowing the agent to visually sense their environment [91]. However, data in a virtual semantic environment is attached to the objects; this can make understanding the virtual world as simple as looking up values in a table. If the agent's model of perception consists solely of looking up values, then the agent exists inside a fully known environment. This may not be plausible for certain cognitive processes (such as agent memory models, decision-making processes, and acoustic information understanding), but can be addressed by modeling various aspects of agent perception. Also, when simulating virtual humans, it is natural to believe that the five human senses will suffice. However, humans augment their senses every day, with tools as simple as thermometers or as complex as radar and night vision systems. Instead of designing all possible senses, an alternative is to generalize what a percept is, and from there, add specifics based on the requirements of a simulation. Determining the best solution for such a problem is generally context dependent, as the information present within the environment plays a necessary role in how the agent can interpret its environment. A simulation author facilitates this process by creating the environment and populating it with information, thus implicitly choosing how an agent understands its world.

An agent can sense the environment, but if it is sufficiently complex, there could be too much information for the agent to realistically process. Humans do not process all objects within an environment at once, and generally only concentrate on a few at a time [91]. For example, a person who glances at a cluttered coffee table may not notice their keys amidst the remotes and books. In order to create more believable behaviors, an agent may need to prune away portions of its sensed search space, keeping only the interesting or stimulating information—thus adding plausible effects of being unaware, unobservant, or overlooking.

Creating an environment that an agent can interact in and understand are related problems. Physical humans not only selectively perceive objects in their environment, but do not always notice details about the objects they do detect. Modelers represent this through levels of detail (LOD); humans generally experience LOD effects with groups of similar objects [94]. To mimic a human's ability to process data, a virtual character would need to have these abilities to exist in and appropriately interact with its environment.

Collections of objects are generally hand clustered by a scene designer, by creating one mesh that appears to be more than one object. Certainly, modelers can represent an aggregate of objects as a single, inseparable model, but doing so can limit a character's interactions with the objects. For example, if the agent needed to only perceive a stack of coins as change, there is little reason for a scenario author to create separable objects. However, if the character wishes to leave a coin from the stack, and take the rest, then it is advantageous for the character to understand both the single coin and stack of coins. While a level designer can encode that semantic information into a stack of objects, the discrepancy between the models and semantic information can cause misunderstandings when being viewed by both physical and virtual characters. By endowing a virtual agent with more human-like abilities, a simulation author may create scenes for virtual humans that display the same information to physical humans.

To allow virtual humans to understand their environment, we have implemented a generalized framework for agent sensing and perception. This framework includes:

- A generalized agent-sensing system that allows simulation authors to create specific senses for a given simulation.

- A perception system that uses heuristics to create sense attention through a bottom-up and top-down process.

- A linear combination of forms of perception that allow for agents to combine perception scores across multiple senses.
- Mechanisms for agents to observe their environment at multiple LODs through object clustering techniques.

13.2 METHODOLOGY

13.2.1 OBJECT AND ACTION REPRESENTATIONS

The representation of objects in our environment is inspired by SmartObjects [115], which contain sets of properties representing semantic information. Semantic properties, p, are categorized into sets, S, and these sets make up the collection of semantic information. Objects may contain one property per set, and may be marked with many different sets. For example, properties p_i and p_j are members of set S_m, and property p_k is a member of set S_n. An object may have either property p_i or p_j, without an effect on its ability to also have p_k. These semantic sets are generally contained globally, and individual objects may inherit a single property from these sets, although the inherited property may change during the simulation. Before the simulation, the sets of semantic properties can be authored by a user, or possibly determined from a commonsense database [166]. A subset of the property sets we are currently using is found in Table 13.1.

Table 13.1: A subset of the property sets understood by our perception system

Property	**Type**
Olfactory type	String set
Visual hue/saturation/brightness	Integer set
Visual luminance	Integer set
Auditory frequency/intensity	Integer set

13.2.2 SENSE PREPROCESSING

In order to endow virtual agents with sense attention, the agent must be able to determine if it can sense an object. We define sensing as the ability an agent uses to determine the presence of a semantically labeled object through some semantic information attached to that object. We represent a change in semantic information upon an object as an event. For example, an agent can sense a pizza through seeing its shape and color or through a pizza's distinctive smell. Different senses should discern different properties of a semantic object, although some properties can manifest themselves in many ways. A semantic object that is hot could have a heat signature semantic information type, detectable by an infrared sense or touch sense. An object that is hot could also glow red and therefore be detected by a visual sense. We consider these two semantic properties distinct, and each one is added to the virtual environment separately.

We provide our agents a method to determine which objects are in their general sensing area. Much like [90], our agents have an area they can sense in, which is determined by the simulation author, and generally varies by sense and by agent. This is determined by two variables, a subtended angle a and a sensing distance d. a and d form polar coordinates that allow agents to determine what objects are within their sensing area, as seen in Fig. 13.2. Unlike techniques such as raytracing, our technique does not inform an agent if an object is blocked by another object. While this removes some realism from certain sensing abilities, we believe that it does not negatively impact the agent. Certain senses, such as auditory and olfactory senses, do not rely on objects being in the agent's line of sight, but being within a certain range of the agent, thus providing them with back up receptors to visibility. (An extended discussion of an agent's auditory sense is given Section 12.)

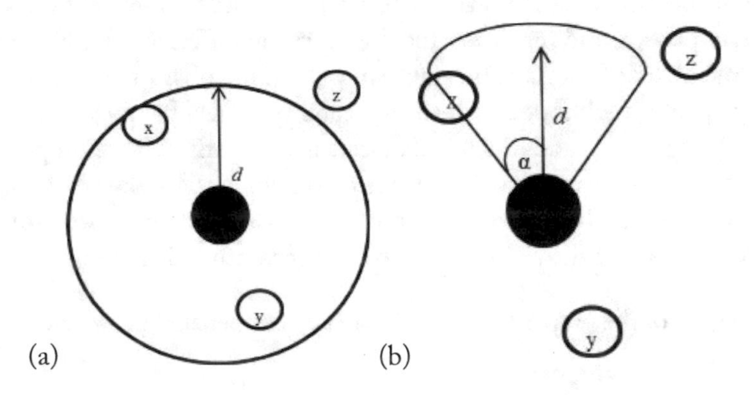

Figure 13.2: Sense preprocessing areas (a) a is 360 degrees around the agent. In this example, x and y are sensed by the agent, but z is not. (b) a is less than 360 degrees. In this example, only object x is sensed.

To determine if an object is in the sensing area, we keep a list of all sorted objects that are then examined by the agent. Objects that exist within this sensing area are fed into the sense where they are pruned from the agent's consideration based upon the semantic properties they contain. At this stage of processing, it is only important for the agent to determine if an object has one semantic property useful to the sense. For example, certain properties, such as a frequency and decibel level, are generally associated with an auditory sense. If an agent is examining an object that has semantic information from one of these two properties, the object is still passed on for further processing. This allows for some generalization between sets and between objects.

13.2.3 SENSING

Physical humans can only process a limited number of items at a time, which is simulated by providing the agent with an upper bound on the working memory portion of the agent system (see Figure 13.1). Only unusual or important information is generally retained, thus rejecting much of what a physical human senses. However, what is generally considered interesting by one sense, when experienced by multiple senses, may not be considered interesting. In order to create more plausible virtual humans, we believe that this ability should also be mimicked. To accomplish this, we model virtual human perception over a multitude of senses.

In order to create specificity when designing perception for multiple senses, we employ a series of heuristics, similar to [259], a subset of which is seen in Table 13.2. However, unlike [259], which determines common attributes from sensed objects, we attempt to rank objects based on pre-authored information readily available from either the agent or the environment (such as the distance between objects or an object's hue). We design heuristics based on two forms of comparison: agent-object interaction, which we label as top-down interactions, and object-object, or bottom-up interactions.

Table 13.2: A sample of the heuristics used in our perception system. Sense is the sense type applicable to the heuristic. Form is whether the heuristic compares objects to other objects (bottom-up) or objects to an agent (top-down). The type of heuristic shows the basic way in which it is calculated.

Name	Sense	Form	Type
Auditory Saliency	Auditory	Bottom-up	Comparative
Olfactory Saliency	Olfactory	Bottom-up	Comparative
Velocity Saliency	Multiple	Bottom-up	Comparative
Interacting	Multiple	Top-down	Selective
Using	Multiple	Top-down	Selective
Useful Object	Multiple	Top-down	Selective

Top-down interaction heuristics allow an agent to create a personal score with the object, therefore yielding different results for different agents. For example, an interacting heuristic, determines if an object (such as another agent) is interacting with the agent. From Table 13.2, it can be seen that we regard all top-down heuristics as selective heuristics, which follow the basic form found in Equation 13.1.

$$h(object, agent) = \begin{cases} score & if\ object\ passes\ selection\ test \\ 0 & otherwise \end{cases} \tag{13.1}$$

Bottom-up heuristics consider object-object interactions. Table 13.2 shows that many of these heuristics use saliency and are comparative. Unlike most systems, we do not use computer vision techniques to create a saliency image, but instead perform comparisons based on the objects through the sense preprocessing phase. While in some cases this will remove information that is generally used in saliency maps, it maintains object-object interaction while removing certain forms of preprocessing (such as background subtraction). Table 13.1 shows that many bottom-up heuristics are comparative, and so the object must be compared to all other objects within the environment. These heuristics, being comparative in nature, also create relative perceptions within the heuristic. A less intense smell, when compared to a stronger one, will be scored much lower using these sorts of heuristics.

Many heuristics used by the agent require statistical processing. Heuristics such as saliency require not only comparisons to each object in the area, but also to the average object score. We implement comparisons to the average, maximum, and minimum score for a set of objects for a given heuristic. Since heuristics can be reused over multiple senses, it is not efficient to embed these statistics within the heuristics, but better to process them after the heuristic has been run on all objects.

After a heuristic, H, is processed for all objects, it is normalized with a weight, ws, over the total of all weights, wn, and added to a hash table object score. After all heuristics are processed for a given sense, that sense is normalized with a weight w and total of all weights n as well. The total score for a given object over all senses is then given as a summation, seen in Equation 13.2. By using linear weights on both heuristic scores and senses, a simulation author can control whether one heuristic or sense should dominate the others. Finally, the hash table is pruned to only the objects that score highest from senses, creating a bounded memory in complex scenes, e.g., as found in [92].

$$u = \sum_{i=1}^{t} \frac{w_i}{n} \sum_{j=1}^{s} \frac{ws_j}{wn} H_j(o) \tag{13.2}$$

13.3 HIERARCHICAL AGGREGATE CLUSTERING

Our agents are meant to perform in complex environments, consisting of hundreds or thousands of objects. However, many of these objects are the same or similar to each other, e.g., as in Fig. 13.3. In many scenarios, related objects are also generally grouped together and the groups can be densely packed. In Fig. 13.3, an agent may find several pens, and most of these are similar blue or black ballpoint pens placed together in a box or next to each other. Considering each of these items individually can be unnecessarily time consuming, and in many cases the agent only cares about the group of pens, and not the individual pen. While modeling a container with pens as a single object is an option, it would mean prohibiting agents from ever using individual pens as a resource. Alternatively, we retain the individual object models and automate dynamic clustering of objects during a perception phase.

13.3.1 ENVIRONMENT-CENTRIC CLUSTERING

The first level of aggregate grouping finds stacks of similar items, and uses pre-selected functions, such as Equations 13.3, 13.4, 13.5, and 13.6 to determine when a set of objects is a group. As this only uses properties of the environment, we consider this level of clustering to be environmentally based. This means that a virtual character can compute this level with a subset of objects in a given sensing range or an object management system may compute this level for all objects in the

Figure 13.3: A real-world cluttered shelf with many unique objects on it.

environment.

$$L_{i,j} = distance(o_i, o_j) \tag{13.3}$$

$$g_{i,j} = \sum_{x=1}^{z} |color(o_{ix}) - color(o_{jx})| \tag{13.4}$$

$$g_{i,j} = (type(o_i) == type(o_j)) \tag{13.5}$$

$$g_{i,j} = (room(o_i) == room(o_j)) \tag{13.6}$$

To start the first level of clustering, an agent receives a set of objects, o, from one of the sensory systems, and attempts to place items into groups based on satisfying combinations of heuristics. While these are not the only semantic properties that objects can be grouped on, all objects are guaranteed to have these properties. For each object o_i in the set of objects o, the agent evaluates each object o_i compared to o_j based on predetermined clustering functions. We have found that testing comparison functions immediately after satisfying binary relationships is superior to testing each function individually. While this is not as general as testing each individually, it is much less computationally complex, creating a sparse matrix rather than several which are dense. Then, using a pre-defined maximum value d, the agent creates a group g from o_i and the minimum valued o_j that is less than d, combining groups as necessary, and can be seen in Equation 13.7. This creates small, similar aggregates of items, such as the clusters made in Fig. 13.4. For semi-static environments, environment-centric grouping can be done as a post-process of loading the graphical environment. Therefore the clustering algorithm need not run in real-time.

$$g = o_j | min(g_{i,j} < d)) \tag{13.7}$$

13.3.2 AGENT-CENTRIC CLUSTERING

An initial clustering provides an environmental understanding of collections of objects. However, many of these clusters are small, containing a few items of the same type. An agent, especially one

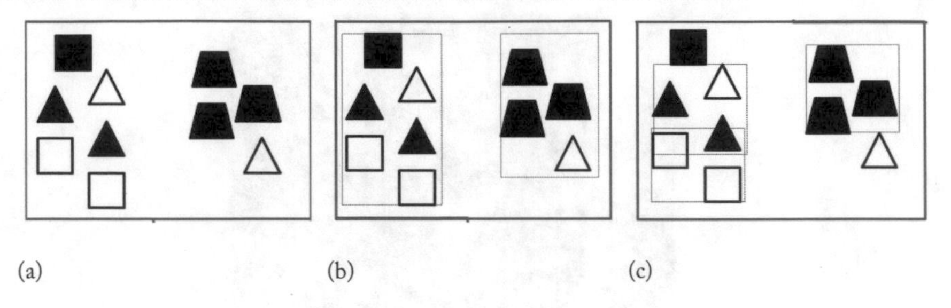

Figure 13.4: A set of objects that can be grouped by their semantics, shown with: (a) no groupings, (b) grouping by proximity, and (c) a combination of type and distance.

at a distance, may not need even this resolution of objects, but can perceive larger and more general collections. For example, an agent may perceive forks, knives, and spoons in a scene. A physical human can see this as either individual objects, groups of forks, knives, and spoons, or as a single group of silverware. Humans have a few ways of creating clusters, by following what is known as the Gestalt laws of grouping [298]. These laws group objects by their *proximity*, *orientation*, or *common fate*, in which the two objects appear as through they will collide given their current orientation. To create a second LOD for the agent, we allow agents to group environmental clusters using Gestalt laws as combinable heuristics.

To calculate larger clusters of objects, a virtual character V is given one or more agent-based heuristics, such as the ones used for the agent's attention mechanism. We have our agents use one of three heuristics: a distance based on subtended angle, seen in Equation 13.8 and Figure 13.5, orientation, seen in Equation 13.9, and common fate, seen in Equation 13.10 heuristic, each

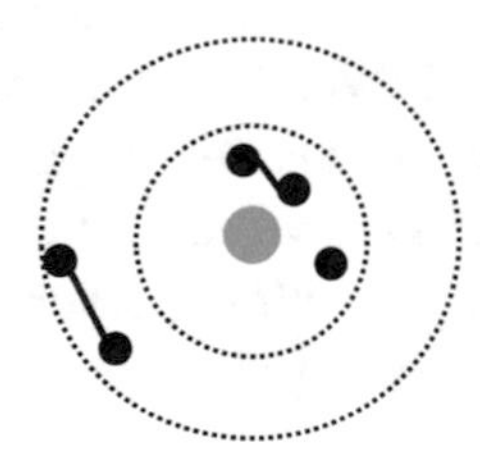

Figure 13.5: A pictorial representation of agent-centric clustering. The further away a set of groups are from the agent, the further away a group of objects may be from each other and still be considered a cluster.

based on Gestalt laws of grouping. Each heuristic measures some angular information between a set of clusters and the character V. We then use Equation 13.7 to determine larger groupings from smaller environmental groupings, using an angle θ instead of a value d, set by the simulation author. It is important to note that for agent-based clustering, the choice of heuristic has an impact on the computational cost of computing such clusters.

$$g_{i,j} = \cos^{-1}\left(\frac{distance(V, g_i)^2 + distance(V, g_j)^2 - distance(g_i, g_j)^2}{(2 * g_i)^2 * distance(V, g_j)^2}\right) \tag{13.8}$$

$$g_{i,j} = \sum_{x=1}^{z} |orientation(o_{ix}) - orientation(o_{jx})| \tag{13.9}$$

$$g_{i,j} = orientation(o_i) - orientation(o_j) \tag{13.10}$$

When comparing groups of objects, very rarely will full equality appear between clusters. Differences between senses may provide more or less objects in a group, and combine groups differently. Instead of equality between groups, we consider a cluster's similarity as an important metric for determining when an agent can consider two groups to be significantly similar. Given two collections of objects, g_i and g_j, an agent determines the percentage of common elements over the total number of items in the two groups, as seen in Equation 13.11. This is known as the Jaccard Index [257]. These two clusters are then considered the same if this generated percentage is greater than a pre-defined threshold t_s, which is generally a high value such as 75%. As groups can contain large numbers of items, this constraint tends to only combine groups with differences of a few items, minimizing semantic differences.

$$\frac{g_i \cap g_j}{g_i \cup g_j} > t_s \tag{13.11}$$

Once similarity has been established, the agent must determine which elements are part of the final collection. We have chosen a least-item principle that preserves the intersection between two groups of items. Most aggregate information that has been previously computed, such as any semantic averaging, must then be recomputed, though aggregate position to the agent is not. This keeps the agent from attempting to recompute the group at every time step.

By using a type hierarchy and similarity to combine groups of low-level aggregate clusters, the agent creates a personalized understanding of their environment. This understanding combines individual object features at potentially long distances, keeping the agent informed of its environment to a certain level. This LOD understanding of the environment frees the agent from processing all the details of a complex environment, and allow the agent to make plausible generalizations.

13.3.3 AGGREGATE PROPERTIES

There are several nuances that must be considered when working with aggregates of objects rather than individual objects. Individual objects within clusters may still be important to a virtual character, so a character must be able to differentiate and reason correctly about both.

Aggregated groups of objects are meant to be considered as a whole object. Therefore, all semantic properties of an individual object must be combined. For properties with a numerical value, an averaging of the data provides sufficient understanding to the agent with the LOD being used. Using the mean is significantly faster than computing mode or standard deviation, and still provides an average understanding of the environment. While information such as standard deviation can be used to determine interesting members of a group, the purpose of aggregate clustering is to consider the group as a whole. Semantic properties can also be members of a set, which cannot be simply averaged together. In this case, we use a majority voting mechanism to determine the most common member of the set for the group. For example, objects in a group can have a *Status* set of properties, with values of *operating*, *broken*, and *idle*. If a majority of the objects have the property *broken*, then the entire cluster is perceived to be *broken*. With both set and numerical properties, properties of the group should not be assigned to properties of the individual. As the objects within a group are still autonomous units, attempting to overwrite properties of a unit with a group defeats the purpose of creating these aggregate clusters at runtime.

In many scenarios, objects in the scene remain static until acted upon by some force or agent, which is generally a player or Non-Player Character (NPC). If an agent exists in a school environment, items such as desks, decorations, and books very rarely move, and generally account for a considerable number of objects in that scenario. These items should maintain their clusters, although an object that has a semantic change, e.g., such as in position or noise, may be added to or removed from a cluster. In these cases, it does not make sense to recompute the environmental clusters for these small changes on a small number of objects, especially given the expensive nature of comparing all items in the whole environment or even a single room. Instead, if the objects that are changed are known, then an object management system partially recomputes clustering information for these items. For example, although most books rarely move, a book removed from a shelf would need to be removed from its environmental cluster and recomputed. This recomputation can be automatically triggered by post-assertions of appropriate actions, such as those found in planning representations [64].

To recompute a set of groups, given a set of objects $\{o_i...o_j\}$ that have had a semantic change, each object must be removed from their containing group g. Then, each $o_k \in \{o_i...o_j\}$ is reclustered by examining the area around each object, assigning each o_k to the first existing group in g that qualifies for all heuristics. If no groups are within the pre-defined distance used to initially group the objects, a new group is added solely containing the moved object.

Finally, there are several advantages of using aggregate clusters with a sensory attention system. Any heuristic that uses relative mean difference must compare each object to all other objects. When the number of groups is much less than that of the objects, then there is a significant reduction in computation time. Agent-based heuristics typically require a specific object, which would generally be a part of a group. These heuristics may then require the agent to search through all objects in all groups, negating any benefit to using clusters. However, our first level of grouping deals only with items of the same type. Many of our agent-based heuristics search for a certain type of object, such as an agent or an object that fits a given task. Therefore, we can use type clustering to prune sections of our objects that would not matter in these agent-based heuristics. For example, if an agent required a *writing implement* to sign its name, any low level group that did not have child of *writing implement* as the first member of the collection would not be considered.

13.4 ANALYSIS AND RESULTS

In order to highlight the difference between using an agent-based linear combination perception system and one using simple selection, a sample scenario has been created. We have modeled a complex diner environment that contains many objects. Situated within this diner is a kitchen, complete with a stove, microwave, pots, pans, cups, food, and other objects typically found in a kitchen. Each of these objects has been labeled with different semantic properties, a subset of which are found in Table 13.1. Certain properties from these sets (such as sound properties) are also added to objects through events. Actuators for the virtual agents are provided through SmartBody [61].

Our scenario has an agent enter the kitchen, examining its surroundings and reporting on the objects it perceives, using vision, auditory, and olfactory senses. All items have at least visual semantic information, and many items, such as the refrigerator or oven, have a sound or smell associated with them. Figure 13.6 displays the agent's observation using a selection method on the left and a linear combination method in the middle. As can be seen from the images, several of the objects are the same. Some items, however, such as the refrigerator, are ignored by the linear combination method. As the refrigerator is only recognized by the auditory sense with both methods, the uninteresting sound that it makes goes unnoticed when more interesting objects, such as the glasses and cups, are in range. The refrigerator's sound is noticed with the selection method due to the lack of objects that have auditory semantics, and that the auditory sense is the last sense to be checked.

We have also developed several scenarios to demonstrate the use of our hierarchical aggregate clustering method with virtual characters. Our virtual characters are placed in a complex (>500 separate objects) virtual world. As above, these agents have been given vision, auditory, and olfactory senses. All of our agents use the same heuristics for a given sense, but heuristics differ appropriately between senses. These heuristics include a distance comparison and agent type ranking for all senses, a color comparison and size ranking for each agent's visual sense, two types of sound ranking heuristics for all agent's auditory sense, and an olfactory intensity ranking for each agent's olfactory sense. Additionally, each agent is given a task-ranking and object interaction heuristic for each sense, for a total of seventeen heuristics for each agent. For cluster-

Figure 13.6: The agent observing its environment. Objects seen in the environment are the objects perceived by the agent. Left: The agent observes its environment using an attention selection method. Middle: The agent observes its environment using our linear combination method. Right: The full kitchen environment.

ing, we use Equations 13.3, 13.5, and 13.6 for environmental clustering and Equation 13.8 for agent-centric clustering.

We use randomly generated environments with object counts between 1,000 and 2,500 to show the effects of clustering on an agent's perception (Fig. 13.7). An agent perceives as much of its environment as possible, using [293] in one scenario and [8] in another. As these environments are very dense, environmental groups tend to be quite large (Fig. 13.8).

Figure 13.7: A randomly generated environment (a) and a pre-authored environment by a modeling expert (b). The agent performs perception and clustering on both environments

In the pre-authored environment, the agent walks into a room that contains several hundred objects. By using clustering, the agent can process more objects of the same type, while still maintaining the Miller principle [187], in which a physical human can only process in short-term memory between five to nine items. This is accomplished because similar objects in the scene are placed very close to each other, such as all the cups and plates on the shelf. Without clustering, the agent only notices one cup or plate at a time, which is not realistic. The agent also is able to more efficiently observe the environment using virtual perception. Using clustering with virtual perception has a computational slowdown of 1.33 compared to the same scene with no perception, and compared to a 2x slowdown when not using clustering.

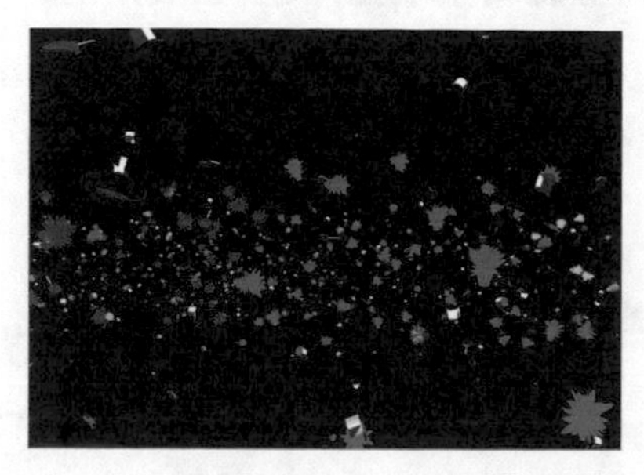

Figure 13.8: The clustering results an agent creates using Equation 13.8. Different clusters are represented as different colors.

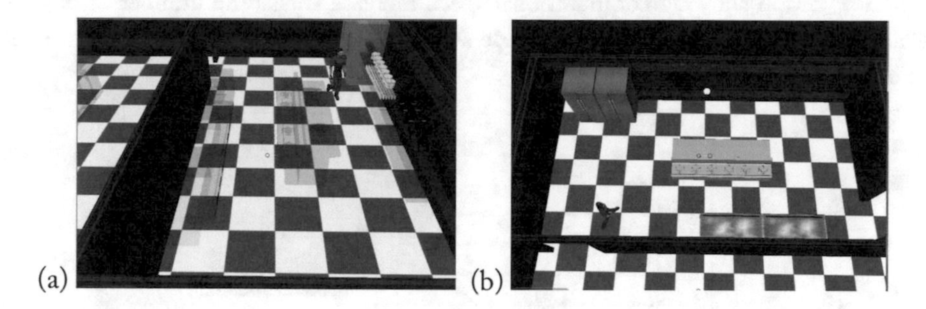

Figure 13.9: An agent perceiving an authored environment, using: (a) our clustering method, and (b) without clustering.

To quantitatively understand the advantages of creating LOD groupings of objects, we compute the computational cost difference when using clustering against not using clustering. This compares the total time for an agent to perceive a scene with a large number of objects. These tests were not performed with rendered objects, just the object meta-data, for all three senses, where every object could be sensed by each sense. As can be seen from Fig. 13.10 and Fig. 13.11, using our clustering system is much faster than having an agent just use a perception system. This is for both having the agent compute environmental clusters at the beginning of each frame, as well as if the environmental clusters are pre-computed. While using pre-computed environmental clusters is faster than using agent-based clustering for one agent, the effect appears to be constant, and there is no observable speedup as the number of objects increases.

We also examine the effects of having to re-compute the environmental clusters at the beginning of each frame (Fig. 13.11). Environmental clustering is apparently advantageous for multiple agents in a system. Using agent-based clustering, the cost of perceiving an environment is approximately the same per agent, which is to be expected. However, when there are multiple agents, the cost of only computing environmental clusters once per frame allows the costs of creating those groups to be spread out over all agents, significantly decreasing the time. From

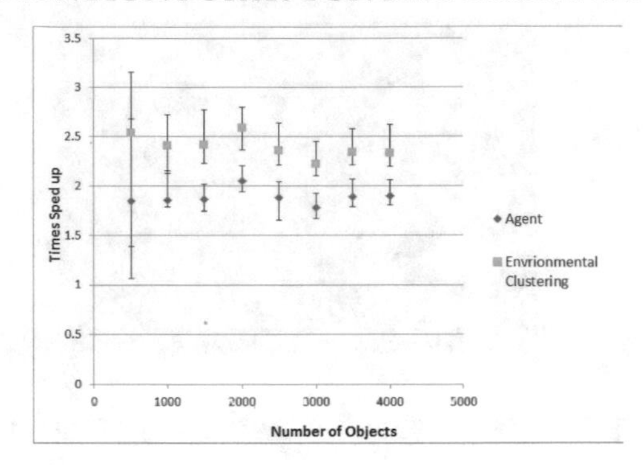

Figure 13.10: Graph showing computational speed increase versus the number of objects in an automatically generated environment for a single agent. The data is compared to the time required to perceive a scene without using clustering.

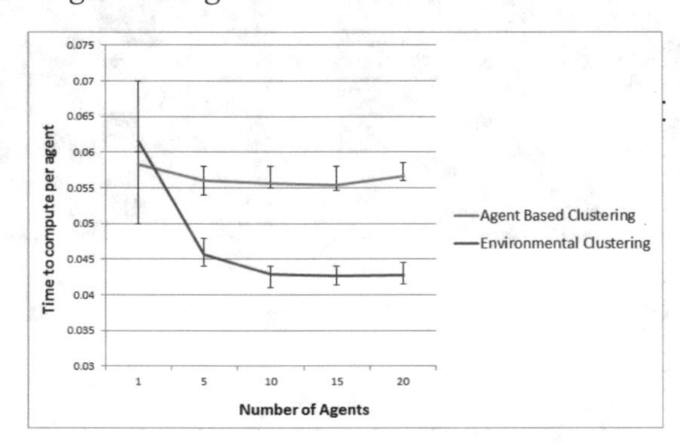

Figure 13.11: The time difference (sec) per agent vs. number of agents for multiple agents to perceive 1,000 objects. The difference for each graph is normalized to one agent.

Fig. 13.11 we see that for a single agent, using agent-based clustering is faster. This is because agent-based clustering only groups objects that an agent can sense. Environmental clustering does not know the parts of the environment an agent can sense *a priori*, and so must compute groups for all objects. However, the computational difference is minuscule, especially compared to the difference when more than one agent is used, so that it is almost always advantageous to use environmental clustering.

To achieve more purposeful, functional crowd simulations, individual agents need to be able to perceive and ultimately respond to their environment in human-like, but computationally feasible ways. The preliminary steps presented in this chapter provide agents with a more comprehensive perception system including several senses and mechanisms to aggregate them for use in decision-making processes.

CHAPTER 14

Semantics in Virtual Environments

Cameron Pelkey and Jan M. Allbeck

An appropriate semantics representation is a key component in the integration of autonomous agent behaviors into a virtual environment [283]. The current processes by which semantics and their respective ontologies are constructed are inadequate for agent simulations. Here we describe a set of tools to facilitate the construction of semantic ontologies and their essential content for use in populated virtual environments.

14.1 INCORPORATING SEMANTICS

When adding object semantics into the virtual environment, there are two primary issues that need to be addressed. The first is to insure the quality of data, so we utilize pre-fabricated, open source lexical corpora. The second is to manage the trade-off between the quantity of information within a smart object and that object's extensibility. Accordingly, we explore the abstraction of smart objects through modularization and runtime-specific attributes.

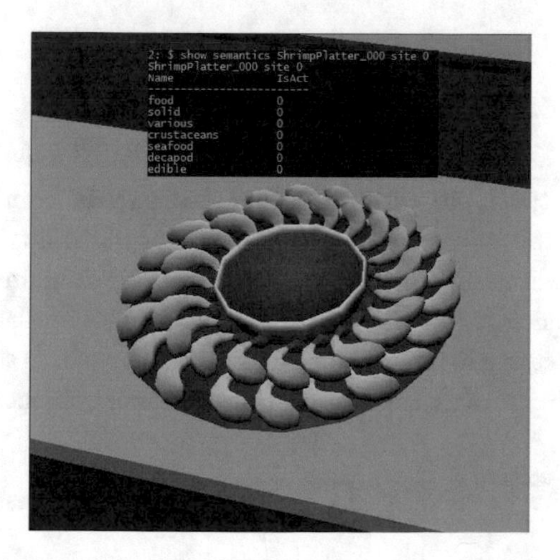

Figure 14.1: A *ShrimpPlatter* virtual object with embedded semantic properties.

14.1.1 LEXICAL DATABASES

The hand-crafting of object hierarchies and semantic ontologies suffers from numerous pitfalls inherent in the manual approach itself. Such systems are often manufactured around specific applications in simulation, resulting in inflexibility in scalability and increased overhead in the time and effort required to maintain the model [283]. Additionally, leaving the simulation authors to construct ontologies arbitrarily may result in models that are, while functional, fundamentally incorrect in their construction, resulting in problems later on.

Relational hierarchies have been developed through crowdsourcing [53, Massachusetts Institute of Technology]. Crowdsourcing by nature operates with an open-world model, under the formal logic assumption that the truth of a statement is independent of whether or not it is known by any one individual to be true [231]. Irrespective of preventative steps taken to ensure the integrity of collected input, the gathering of information from such a diverse population results in data that may be misleading, irrelevant, or otherwise incorrect. This approach is fundamentally in conflict with the closed-world model usually adopted for games and simulations, where what is known is fact and what is not known is assumed false. Using crowdsourced data has severe drawbacks that make it highly impractical for use within the context of a closed-world system.

To ensure a more consistent means of reliable data collection, we instead move toward the use of established lexical corpora such as WordNet [Princeton University], VerbNet [Palmer and Kipper], and PropBank [Palmer]. These corpora enjoy freedom of availability and ease of accessibility. Additionally, for each corpora, there exist a variety of development tools, facilitating not only the manipulation of corpora data but also its incorporation into a development effort. Finally, the corpora are set up such that the information contained within any one corpora may be used in conjunction with the other two, which facilitates creating more complex, complete environments.

14.1.2 MODULARIZED SMART OBJECTS

As originally shown by Kallmann and Thallmann [116] and Douville et al. [49], there exists a disjunction between the specificity of information that may be contained within a smart object and the extensibility of that object with respect to its applicability beyond the given scenario. In order to allow for the inclusion of increasingly detailed, domain-specific information in the construction of smart objects, while maintaining the concept's fundamental notions of generality and extensibility, it is necessary to revise the approach itself by which smart objects are traditionally constructed by introducing further abstraction through smart object modularization.

14.2 SEMANTIC GENERATION

The system is built in two parts. The pre-processed component comprises two tools for semi-automating the generation of both an object hierarchy and the requisite semantic modules for that hierarchy. The runtime component is the Semantic Module Handler, which acts as abstracted knowledge layer between agents and objects for managing the loading, dissemination, and interaction with all semantics in an environment (Fig. 14.2).

Figure 14.2: The agent interacts with the virtual environment with the assistance of semantic affordances.

14.2.1 HIERARCHY GENERATION

All pre-processing components are built in Python [294]. Object hierarchies are generated using the WordNet corpora via the Natural Language Processing Toolkit [21]. The WordNet database is organized into a hierarchy of *synsets* (synonym sets) of the form *word.PartOfSpeech.synsetNumber* (e.g., *rabbit.n.01*). Each synset is representative of a single meaning of a given word; words may have multiple definitions. The WordNet hierarchy is a directed acyclic graph with all nodes extending from the most general form: *entity.n.01*. Therefore, the challenge in constructing hierarchies from WordNet comes from first associating each object in a scene to its proper synset definition, then properly trimming the resulting hierarchy into a tree structure more suitable for use in simulation (Fig. 14.3).

To construct the object hierarchy, the simulation author just supplies a list of all thc unique objects to be represented within a scene, labeled in the form *object[: keywords]*.

The keyword arguments assist in determining the proper synset to be associated with the given object. Although optional, if none are provided, the author may be prompted to select a specific definition during hierarchy generation. For example, if an environment is to include a number of common household pets, the author may include the following:

rabbit : fur ears lagomorph
cat
dog : canine mammal friend
An example hierarchy generated using this method is presented in Fig. 14.3.

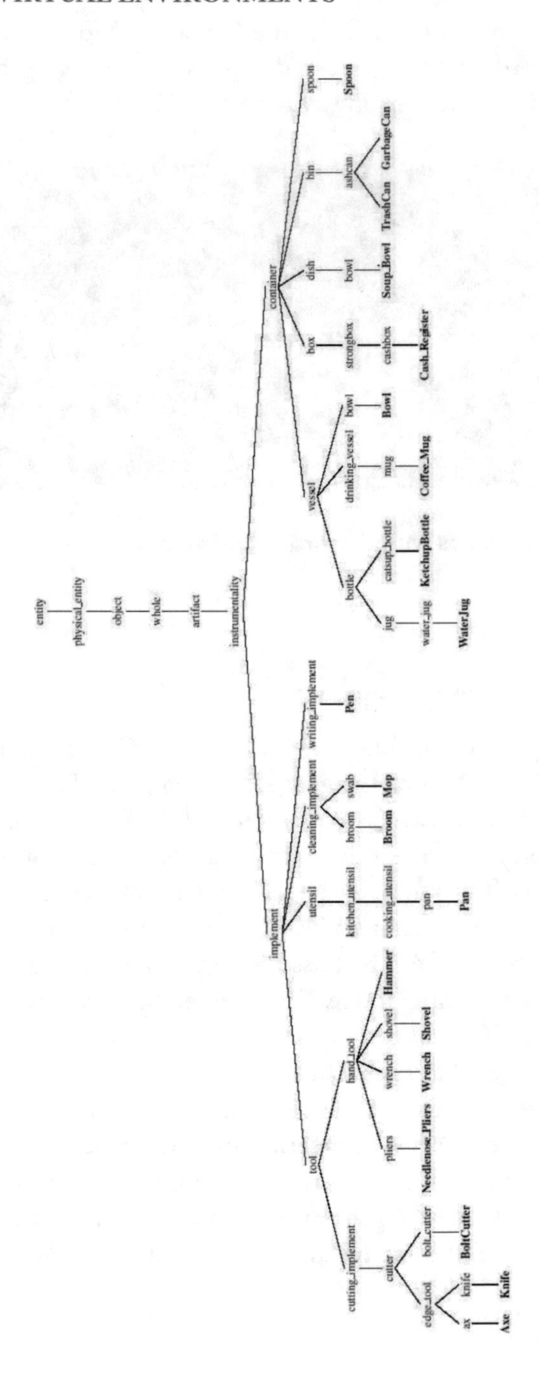

Figure 14.3: A sample object hierarchy as generated through WordNet.

Once a hierarchy is generated, a MySQL table is injected into the simulation database ready for immediate use by the author. At this point all appropriate semantic modules are also generated and the object-module associations are stored within the database.

Tests were performed by creating object groupings of various sizes, consisting of objects drawn from randomly extracted WordNet synsets. For each of these groupings, the items were organized into a list in the *object : keywords* format and pre-processing tool used this list to generate a hierarchy. The resulting object-synset pairs were then compared against the original selection of objects and the synsets from which they were generated. The overall accuracy of the system was determined by the rate of successfully matched object-synset pairs.

Table 14.1: Testing of the hierarchy generation process was conducted over an increasing number of randomly generated objects, measuring the degree to which synsets were correctly mapped back

Objects	Found	Mismatch	Accuracy(%)	Time(sec)
50	49	0	98.00	12.97
100	100	5	95.00	18.70
150	150	5	96.67	24.83
200	198	4	97.00	33.02
250	249	3	98.40	37.06
300	299	8	97.00	42.07
350	349	5	98.29	55.16
400	400	11	97.25	62.19
450	447	11	96.89	67.98
500	498	6	98.40	69.96

This generation method is extremely capable of accurately constructing object hierarchies, with an average successful match rate of 97.29%, as shown in Table 14.1. The time shown includes not only the construction of the hierarchy but also the generation of all related semantics modules and the writing of all necessary information to the database. For every 50 objects, there is an average 6.33 second increase in overhead (Fig. 14.4). There is no need to rerun the database generation tool unless modifications to the hierarchy itself are needed.

There is no guarantee that the linguistic content of WordNet will be optimal for all applications. When compared against other knowledge resources, however, such as ConceptNet or OpenCyc [42], the WordNet framework provides a means for quickly constructing hierarchical structures that may be later augmented by these resources. Databases such as VerbNet and OpenCyc also provide relative links within their linguistic models to the WordNet corpora, facilitating such future expansion.

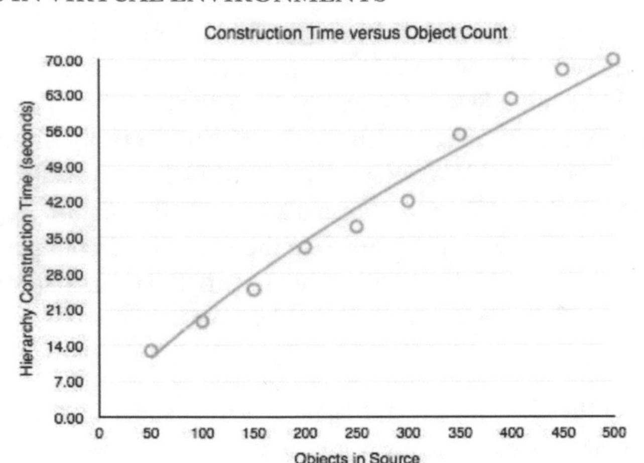

Figure 14.4: Increasing the number of unique objects in a scene has minimal effect on the time taken to compute a new hierarchy and generate the appropriate semantic modules.

14.2.2 SEMANTIC MODULARIZATION

Modularizing smart object components facilitates the construction of information both during the environment design process as well as at runtime. An XML-like schema is utilized, as shown in Fig. 14.5.

This file structure is easy to maintain and extend by the simulation author. The name of the file corresponds to the specific WordNet synset from which it was produced to avoid conflict with similarly defined objects. Semantic affordances are stored as *qualitative* properties in the form <*type value*>. This allows further delineation of the semantics into textual descriptors relating to higher-level illustrative properties and those relating more directly to specific actions that may be afforded by the object.

Modules themselves are not limited to establishing attributes of specific objects as the example illustrates, but may be defined instead as collections of related affordances of any general type. As a result, it is possible to have modules such as *Dry*, *Mammal*, or *John*.

To specify the relation of modules to their respective virtual objects at runtime, these modules are stored in an indexed list within the runtime database. Correspondences between *object* and *module* are then stored in a separate table, which is then referred to when loading objects into an environment.

14.2.3 RUNTIME PERFORMANCE

The management of semantic modules at runtime is controlled by a separate handler, the *Semantic Module Handler*, in an implementation similar to the Actionary as developed by Bindiganavale

```
bowl_01.mod
<name>bowl_01</name>
<modType>object</modType>
<class>PhysicalObject</class>
<state>entity</state>
<qualitative>
    <prop val="round"/>
    <prop val="vessel"/>
    <prop val="holding"/>
    <prop val="top"/>
    <prop val="food"/>
    <prop val="container"/>
    <act val="hold"/>
    <act val="contain"/>
</qualitative>
```

Figure 14.5: A sample semantic module containing relational semantics for object *bowl*.

et al. [20]. A hierarchical approach was necessary to allow for a sufficient level of precision and control over the dissemination of semantics across an increasingly complex virtual environment.

Modules are loaded into the simulation once on an as-needed basis, utilizing the open source tool RapidXML [110] to limit the overhead cost to fractions of a second per file. As each file is loaded, a corresponding container, an SMod, is created which stores the appropriate semantics to be distributed to further instances of the object without necessitating subsequent reloading. Instances of *smart objects* are appended with *site* containers, which act as intermediaries between the object and handler. The site containers store semantic regions, which are capable of representing individual components or the entirety of an object. Semantics from the handler are transferred into these regions for subsequently interactions. The complete system hierarchy is illustrated in Fig. 14.6.

Given a hierarchy constructed of 50 objects chosen at random, with a resulting 300 generated semantic values, system scalability was measured over a set of randomly generated environments. The two primary metrics are the initial overhead of injecting semantics into the environment (Table 14.2) and the processing time required for semantic-based agent–object interaction (Table 14.3).

The processes of loading an environment and injecting it with the necessary semantics occur in parallel. Therefore, in order to measure the speed at which semantic affordances were both loaded into the system and disseminated across all appropriate virtual objects, it was necessary to first calculate the average time required to load each scene without the given affordances. From

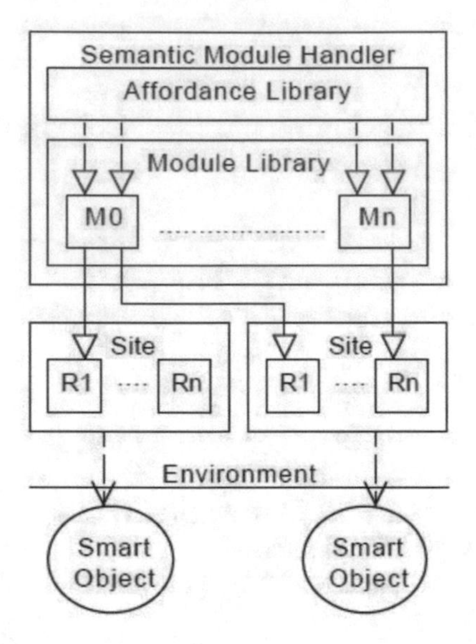

Figure 14.6: An overview of the semantic injection system hierarchy.

Table 14.2: Performance of the distribution of semantic affordances across a suite of increasingly large, randomly generated environments

Objects	Approx. Semantic Count	Injection Speed (sec)	Total Load Time (sec)
100	600	13	36
200	1200	14	37
300	1800	17	40
400	2400	18	41
500	3000	21	44

this, we were able to approximate the injection speed of semantics within each environment, as shown in Table 14.2. Overall, the semantics injection speed increased at a rate proportional to 2 seconds per 100 objects or 600 semantic inclusions (Fig. 14.7).

While the pre-process overhead time required for injection is demonstrably linear with respect to the size of the environment, it must also be computationally feasible to utilize these semantics during runtime, particularly through the execution of agent tasks. To test this, the agent was placed within the generated environments and directed to walk from one end to the other, such that all objects were required to enter its field of perception. The complexity of agent search through its environment (Table 14.3) is roughly $O(\log(n))$ with respect to the number of searchable semantics.

Table 14.3: The process of searching through object semantics within an agent's field of perception has a negligible impact on the overall performance of the simulation

Objects	Approx. Semantic Count	Average Measured Search Time (ms)
100	600	1
200	1200	2
300	1800	2
400	2400	3
500	3000	3

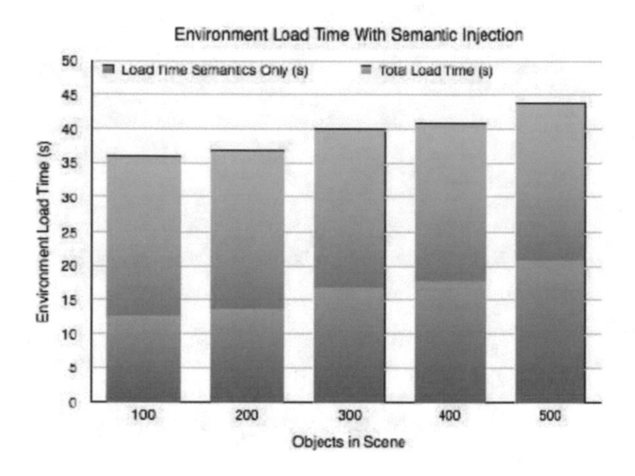

Figure 14.7: Average time necessary to inject a virtual environment with semantic affordances, proportional to the total load time of the scene.

14.3 LIMITATIONS

We have presented a method geared toward facilitating the construction of semantic ontologies for virtual environments, as well as an approach for the generation of semantic affordances and their inclusion within these environments. The method has been shown to be computationally efficient in its implementation and scalable across diverse environments of varying size and semantic inclusion.

Limitations of the current system concern ambiguities that are present in both object identification and in generated semantics. Object polysemy, the coexistence of many possible definitions for a given object, and the under-specification of an object during hierarchy construction may result in misidentification, thereby associating an object with incorrect affordances. Semantic ambiguity results from the generation of an excess number of affordances that provide little-to-no useable information to the agent. Human intervention is required in mitigating the results of either forms of ambiguity; however, the automated process still lessens the workload remaining for the simulation author.

Semantic ambiguity is also present in the generation of synonymous semantic affordances, increasing the search space and obfuscating the agent decision-making process. The process of reducing this redundancy is the subject of future research. A future extension would be to more fully integrate the framework with high-level agent planning systems, using the embedded semantics directly to assist behavior and decision-making processes.

CHAPTER 15

Conclusion

We have integrated sound propagation and human-like perception into virtual human simulations. While sound propagation and synthesis have been explored in computer graphics, and there exist extensive studies on auditory perception in psychology, ours is the first work to enable virtual humans to plausibly hear, listen, and react to auditory triggers. To achieve this goal, we have developed a minimal, yet sufficient sound representation that captures the acoustic features necessary for human perception, designed an efficient sound propagation framework that accurately simulates sound degradation, and used hierarchical clustering analysis to model approximate human-like perceptions using sound categories.

Our method is not intended for auralization [235], but serves as a companion to auralization—enabling virtual agents to hear and classify sounds much like their real human counterparts. Auralization is not a prerequisite for this capability. Moreover, we need to compare the simplified SPR method with other forms of data representations for different types of sounds, as described in [40]. There are no technical barriers to extending the TLM algorithm into 3D but we have no obvious reason to do so for the envisioned environments and situations.

Once agents have both visual and auditory senses, they need to adapt any decision-making to the multiplicity of possible inputs. We have provided a method to create perceptual attention based on a combination of different senses. This is accomplished using a preprocessing step to determine objects and events capable of being considered by a given sense, and a ranking step to determine objects that are useful or interesting to the agent. This provides the agent with an ability to differentiate objects in its environment through the use of all available information.

Future research will examine the heuristics used in this work. Various heuristics may be applicable only for certain situations, and so the ability for the agent to adapt and control its heuristics may provide faster and more interesting results. Optimizations of these heuristics, especially the comparative ones, would also prove useful. An exciting avenue of future exploration is to leverage SPREAD for sound localization [299] and speech perception, and integrate that into existing virtual human behavior frameworks.

To promote meaningful agent-object interactions, we have also introduced methods for automating the semantic labeling of objects in virtual worlds from lexical databases. The methods presented create a more generalized, robust object ontology and that can be used in agent decision-making processes and narrative generation like those discussed later in this volume. Future work in this area would include generating action ontologies through similar means and connecting the action and object ontologies to obtain object operational information (e.g., to eat, an agent needs food).

PART IV

Agent-Object Interactions and Crowd Heterogeneity

CHAPTER 16

Background

Real human crowds and populations are not uniform. Not everyone performs the same behaviors and even similar behaviors are not performed in the same ways [126], and subtle movement differences denote widely different personalities [196]. Physical human populations are heterogeneous, so virtual populations should reflect that. Individual differences include psychological factors, such as emotional state and personalities. Agents' physical and psychological needs can also influence their behaviors and create differences between agents. Assigning agents roles can also create variations in behaviors and provide a richer depiction of a population. Previous agent experiences can also influence behavior and create variations. An agent might remember a shortcut to the market or have a memory of where to find the key to the storage closet that another agent does not. In this part, we will look at some ways create variation in virtual crowds.

Memory systems, appearing to be an inevitable component in intelligent agent architectures, have been studied and developed from a variety of points of view. In animated, autonomous agents research, several groups have combined vision and memory systems to allow virtual characters to perform navigation and path planning tasks [197, 274]. Others have used memory to facilitate and enhance agent-object interactions [147, 215], crowd simulations [209], believability and intelligence in synthetic characters [29, 164], and virtual actors in dramas [178]. In these efforts, memory was not the main focus. It was a part of larger agent architectures that also included other components such as perceptual processing, dialogue units, action selection modules, and goal and plan generation systems. Given the overall complexity, the memory model is often treated simply as permanent or temporary information storage with relatively simple structure, and is not intended to be scrupulously designed to achieve human-like performances in games and other applications.

More specifically, a great deal of work has explored using certain types of memory found in the psychology literature. Episodic memory [281, 300] collects individual experiences that occurred at particular times and locations. Systems using episodic memory include: a pedagogical agent named Steve [227] who can explain its decision-making process for various task steps, Brom et al. [27] use it to achieve longer duration storytelling activity in a gaming environment, and Gomes et al. [77] apply episodic memory to emotion appraisal. Similar to episodic memory but with broader scope, autobiographical memory has also been incorporated into many applications: e.g., Dias et al. [44] uses it to enable synthetic agents to report their past experiences, and in [93] actors (similar to the agent Steve) can tell stories over a limited time period based on their autobiographical memory. While significant results were achieved, these efforts focused primarily on retrieving episodic knowledge and had memory modules crafted for specific tasks such as storytelling, communicating, and social companionship [169].

Cognitive computing research is also relevant to agent behaviors. A memory model is used for cognitive robots [46] and autonomous agents [33, 43, 67]. Considerable effort has been expended on cognitive architectures such as SOAR [152], ACT-R [4], and CLARION [267]. These architectures generally have relatively comprehensive memory models in their large software infrastructures and are capable of simulating many psychological activities. For example, Nuxoll [198] attempts to build an event-independent episodic memory system integrated with SOAR. However, there are some issues with these cognitive architectures. Generality is pursued at the loss of some specific elements. Consider SOAR, one of the virtual and gaming environment-friendly cognitive architectures [151, 152]: while its usage has been demonstrated in several gaming applications, most of them take place in simple environments with agents capable of conducting only a limited number of behaviors and interactions with the environment. In addition, using these cognitive architectures requires considerable effort and seasoned programming skills. In contrast, a parameterized model with tunable values could be easily adopted by users to endow virtual characters with heterogeneous features.

The work presented in this chapter attempts to fill some gaps in mapping cognitive memory models to animated agents. Toward that end, we create a synthetic parameterized memory model based on lessons learned in the cognitive computing and agents communities. Our model is grounded in a game world with autonomous virtual humans conducting life-like interactions with objects and each other.

Although memories can provide continuity in behaviors and give preferences for locations and objects in interactions, they do not readily inform how agents interact with other agents. Such relationships are crucial for crowd simulations. Two or three actors could interact in complex ways, a group of actors might participate in an event, or a large crowd with hundreds and thousands of actors could exhibit an emergent global characteristic. For example, a user may want to author huge armies in movies, the repercussions of a car accident in a busy city street, the reactions of a crowd to a disturbance, a virtual marketplace with buyers and vendors haggling for prices, and thieves that are on the lookout for stealing opportunities. The existing baseline for multi-actor simulations consists of numerous relatively independent walking pedestrians. While their visual appearances may be quite varied, their behavioral repertoire is not, and their interactions are generally limited to attention control and collision avoidance. The next generation of interactive virtual world applications require functional, purposeful, heterogeneous actors with individual personalities, roles, and desires, while exhibiting complex group interactions, and conforming to global narrative constraints.

In this chapter we present a method to capture social crowd dynamics by mapping low-level simulation parameters to the OCEAN personality model. Each personality trait is associated with nominal behaviors—facilitating a plausible mapping of personality traits to existing behavior types. We validate our mapping by conducting a user study which assesses the perception of personality traits in a variety of crowd simulations demonstrating these behaviors [51]. Finally, we also include a method to incorporate social roles and needs into a virtual population, giving the characters more purposeful behaviors [163].

CHAPTER 17

Parameterized Memory Models

Weizi Li, John T. Balint, and Jan M. Allbeck

Non-Player Characters (NPCs) are autonomous agents in game worlds who portray enemies, allies, and neutral characters. Over the past three decades, there has been a great deal of work on improving NPCs. However, while a majority of it has gone into creating more visually appealing and animated characters, development of the underlying intelligence for these characters has remained fairly stagnant, creating strange and undesirable phenomenon such as repetitive behaviors and a lack of learning and knowledge understanding. A character may appear as a photo-realistic knight in shining armor, but can only greet the player or fight with the player monotonously. This lack of depth diminishes NPC believability and creates a less enjoyable gaming experience for the player.

There are many different forms and functions that NPCs need to fulfill, and these commonly correspond to their roles, relevance, and importance to a player. In many of these cases, when the purpose of agents is to stay transiently and blend into the environment, it's less likely a player would spend a great deal of time examining each one of them. For example, if there is a squad of enemy soldiers the player must fight, it is doubtful that the player will spend considerable time monitoring each individual soldier's behaviors. In these contexts, there is not a great need for strong AI techniques, and it is generally impractical to implement and execute such techniques for a large group of agents. However, for those agents meant to serve as companions or enemies that are central to the game story and thus exist for longer periods of time, the player will most likely spend a lot of time interacting with them and also observing their behaviors. For these characters, techniques that create more believable agents are needed, and these types of NPCs are considered here.

While there are many ways of improving the believability of virtual agents, in particular we see a human-like memory model helping to achieve this goal in two ways. First, given that information storage is inevitably needed for agents to reason and interact with their environments, a human-like memory model that includes false memory and memory distortion can result in more human-like performances. These human-level abilities (i.e., neither sub-human nor super-human) will provide more plausible interactions with the player, including more reasonable competition. Secondly, an event-independent memory model can make longer simulation more plausible and allow agents to carry their knowledge to other scenarios without massive editing. To realize such a system, certain aspects of human memory are necessary such as learning, forgetting, and false memories. While there are many techniques, such as scripting and behavior trees, that can be used to simulate an agent's memory model and can create the illusion of false memories and forgetful agents, these techniques usually require a great deal of crafting by a game play author in order to create some form of believability. A memory model that causes an agent to forget or create hazy memories provides the virtual character some variability in its understanding of the world and what has taken place. This variability is inherent within a reasonable human-like memory model, and these differences do not have to be enumerated and explicitly written by the game author, as they would with prior techniques.

Toward this end, we have developed a memory model that includes components for *Sensory Memory*, *Working Memory*, and *Long-term Memory* and have designed it to contain features such as forgetting and false memories. We also provided multiple parameters related to various memory capabilities. These parameters can be set by level designers or game systems to vary the difficulty level of games. To determine a reasonable range of values for these parameters, we conducted a user study in a 3D game environment and carried out a performance analysis.

We will define an agent memory model that supports a variety of human-like psychological activities such as forgetting and false memories, offers flexibility through user-tunable parameters, and is grounded in a 3D environment where agents are capable of demonstrating complex, emergent, and social behaviors.

17.1 MEMORY SYSTEM

We will first explain the memory representation and then detail its components which include *Sensory Memory*, *Working Memory*, and *Long-term Memory*.

17.1.1 MEMORY REPRESENTATION

Our memory model is a directed graph with nodes representing concepts and edges linking concepts. Edges are directed following the observation that humans generally archive memories in a certain order, and this order does not necessarily work in reverse. This representation can benefit NPCs, such as in a scenario where an NPC has been given a formula of a medicine which can be made only by adding its constituent elements in a certain sequence. While some elements of this formula may be forgotten, the agent should still be able to create and preserve a sense of sequence in order to facilitate resolving the missing elements. An example of this would be an agent thinking, "I remember that in order to make this medicine I need one more item after I use the purple potion, but I forgot which item that is. I should search in this ancient book to determine what the item after the purple potion is." Without directed edges between between concepts, this activity is harder to capture.

In addition, we have implemented a strength factor for both nodes and edges, $Node_{strength}$ and $Edge_{strength}$, respectively. $Node_{strength}$ indicates how strongly a concept is encoded within the memory model while $Edge_{strength}$ denotes the relative ease of going from one node to another. These strength factors add believability to characters; specifically, while many objects afford a set of actions, an agent should select not a random object, but the most familiar object for a particular task. For example, if an NPC is trying to build a birdhouse, it is more believable that the NPC uses its frequently used hammer over a nail gun. While both items can perform the same tasks, the NPC is more accustomed to using his hammer, and thus should be more likely to do so. Currently, both $Node_{strength}$ and $Edge_{strength}$ share the same integer range, 1 to 10, and this range is further divided into two stages, a strong and a weak stage. By creating this division, the model is able to support memory distortion. The division between the two stages is chosen by the user, who does so by choosing the threshold values $Node_{threshold}$ and $Edge_{threshold}$. Therefore, if someone has chosen $Node_{threshold}$ to be 5, then a value of 1 to 4 would create a weak stage node and values between 5 to 10 would be the node's strong stage. While the threshold may be different, the same logic applies to edge values. When nodes and edges are in weak stage, the phenomenon known as false memory can occur with probability $P_{false} = 1 - \frac{Node/Edge_{strength}}{Node/Edge_{threshold}}$. So, if $Node_{threshold}$ has been set as 5, then nodes in the weak stage with values from 1 to 4 have corresponding probabilities from 80% to 20% to be forgotten. This simple design allows us to simulate partial memory fuzziness. Currently instead of having a sophisticated mechanism for selecting an incorrect node and edge to replace the correct ones during the course of false memory generation, the false concept will be picked among neighboring concepts based on an object ontology of the environment.

17.1.2 SENSORY MEMORY

One module of our memory model is called *Sensory Memory*. This component maintains transient information captured by the sensory system. In this discussion we only address vision. Psychology studies have shown that information coming from the environment does not contact memory directly. Instead, visual information spends a short time in an interface between perception and memory [6, 236]. This interface is always present in humans, and its presence is evidenced in simple phenomena such as the light streak observed when swinging an illuminating object swiftly in the dark. We believe this module is also necessary due to the fact that humans never record complete events [188]. Instead, key elements are recorded and later, with the aid of environmental cues, the elements are used to reconstruct past scenarios and events. Here we use $SM_{capacity}$ to indicate how many cues will be potentially maintained in the sensory memory module. Inspired by findings in [187], which determined that a human can process 5 to 9 items at once, and considering the design choice that all elements from sensory memory will transit to working memory with various strength factors, we have chosen the range of $SM_{capacity}$ as 1 to 10. This implementation helps preserve *Sensory Honesty* which has been argued to play a vital role for synthetic characters [29]; essentially it provides a level playing field as the human player and the NPC would have similar sensory processing rates.

17.1.3 WORKING MEMORY

We also developed a *Working Memory* module which is seen in many virtual agent architectures. Working memory stores information currently being used by the agent for a variety of cognitive activities such as reasoning, thinking, and comprehending. Studies have shown that in order to use declarative long-term memory elements, one has to extract the material into working memory first. The working memory has been documented to have much smaller size and shorter information-retaining time compared to long-term memory [6]. From this, we (and other memory architectures, such as SOAR) decide that only one graph structure could exist in the working memory at a time. In addition, while one graph could contain multiple nodes and edges, the actual reinforcement rate on each node and edge depends on the total number of nodes and edges and the information linger time. The equation for calculating the reinforcement rate is: $Node/Edge_{rate} = \frac{Informationlingertime(secs)}{Totalnumberofnodes/edges} * \frac{WM_{scale}}{10}$ where WM_{scale} shares the same range (1 to 10) with the strength factor. Therefore, if the working memory currently contains 4 nodes and 3 edges and this information resides for 12 seconds while WM_{scale} equals to 10, then each node and edge will get its strength factor increased by 3 and 4 accordingly. The interpretation of this procedure is that a few items lingering for a long time in working memory would have their strength factor values higher than those of many items lingering for a short time. This design decision is inspired by findings in [260] which found that memory reactive time increases in a linear fashion when the number of items in the memory increases. Although we do not associate a reactive time with each node and edge, we believe the strength factor can act as an indicator manifesting memory retrieval difficulties. A user can control the amount of information being processed in the working memory, achieving a similar effect to setting a working memory duration and processing strategy without concerns about individual memory differences and the graphical complexity of specific scenarios.

17.1.4 LONG-TERM MEMORY

The final module is the *Long-term Memory* module; intuitively, it maintains an extensive set of concepts for a long duration. Unlike working memory, long-term memory can contains multiple graph structures. While both $Node_{strength}$ and $Edge_{strength}$ will only get strengthened in the working memory, in long-term memory they suffer from decay. Currently, the decaying activity occurs every 5 minutes and the decaying percentage roughly follows the classic Ebbinghaus forgetting curve [52]. While there is debate within the scientific community on whether memory elements are completely forgotten, from the engineering perspective we have decided to remove (forget) any node and edge with $Node/Edge_{strength}$ below 1 and consider any element with $Node/Edge_{strength}$ higher than 10 as permanent.

The overall memory process is illustrated in Fig. 17.1. Through perception, certain environmental cues are passed into the sensory memory. If the number of cues in a place are greater than $SM_{capacity}$, the cues will be randomly selected. These cues are then sent to the working memory, where they are formed into a strong connected graph with a minimum value of $Node/Edge_{strength}$. If a concept has additional properties such as color and material, only the concept and its properties would be linked together. Next, cues will be matched against long-term memory and elements (i.e., nodes and edges) above $Node_{threshold}$ and $Edge_{threshold}$

will be retrieved correctly while elements below this threshold would be subject to potential concept replacement. For example, if the correct concept "Coin 0" has a strength factor below $Node_{threshold}$, then it is possible it will be replaced by "Coin 1" under the meta-concept "Coin" (only if "Coin 1" exists, otherwise it might be forgotten). In other words, a different coin might be incorrectly remembered. Currently, we only consider replacing concepts that are at the same level of the correct concept in our object ontology. This design decision is supported by the data we collected through a user study in which we found that most false memories involved players picking an object from within the same meta-concept instead of completely unrelated concepts. After successfully constructing a single graph structure in working memory, reinforcement will start according to the total number of nodes/edges, information linger time, and the value of WM_{scale}. Finally, memory material will transit to long-term memory with their various updated strength factors.

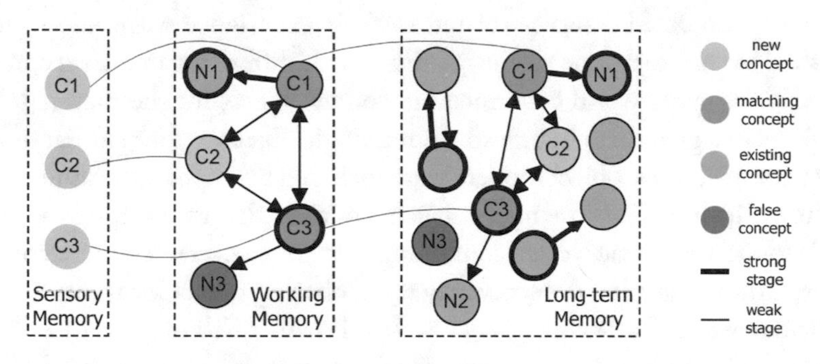

Figure 17.1: The general working pattern of sensory memory, working memory and long-term memory. "C" stands for Cue and "N" stands for Node (not all nodes are labeled).

17.2 EXAMPLE AND ANALYSIS

We implemented a game (Fig. 17.2) in which heroes are tasked with saving their princess, who is locked away in a tower. The first person to find the magic gems to unlock the door and free her will win her everlasting love. Players (both human and an NPC we named Carl) begin by exploring the environment to find two magic gems: one red, one blue. As they explore different areas, they find many objects, but no magic gems. After a time, they encounter a villager NPC holding a red gem. Interacting with the villager, they discover he would be willing to trade the red gem for an iris. Players then have to try to remember where they might have seen an iris. If they cannot remember, they start looking around until they find it. While looking around, their memories of the environment are reinforced. A similar procedure is followed for the blue gem.

While our NPC Carl can certainly successfully complete his task, we are more interested in how his actions can approximate real human performance, and what memory model parameters would be appropriate for different human player skill levels. To explore this we conducted a user

Figure 17.2: (a) The game environment. (b) Carl is exploring the environment, trying to find the desired gems. (c) Carl talks to a civilian about trading his gem. (d) Carl uses his memory successfully to find the required item for trading.

study. In total 31 subjects (15 female, 16 male) participated. Before playing, subjects took a simple memory test and completed a survey related to their experience with video games. Then each subject was asked to play the game solo eight times with different game level complexities. In the first four rounds, the game world contained only eight objects for the players to remember. The number of objects in a given area increased from a single object to four similar objects of different colors. In later rounds, more object models were included as opposed to differentiating by color. Results are shown in Fig. 17.3 in which subjects are classified as having good, medium, or bad memories. We found that game worlds containing more objects created more confusion, resulting in players more often forgetting or incorrectly remembering object locations.

In particular, we set Carl's $SM_{capacity} = 7$ and $Node/Edge_{threshold} = 5$. With the gathered data, we were able to tune the memory model parameters to make our NPC achieve human-like performances. By scaling WM_{scale}, which determines the reinforcement rate on nodes and edges in the working memory, we found when $WM_{scale} \geq 8$ Carl achieves similar performances to human players with good memory. When $5 \leq WM_{scale} < 8$ medium memory performance is obtained and when $WM_{scale} < 5$ performance of a human player with bad memory is reached. The data provided some other interesting findings:

- Using all different object models has no significant improvement over using limited models with different colors (which was assumed to be more confusing) in terms of recalling their locations. This implies that creating more models in games may have limited utility over color cues in helping players remember their locations.
- When the total number of objects in the environment was more than 20, 90% of players forgot the desired item's location even when the item was among the last 6 items seen. This indicates that reinforcement is not strictly distributed to the latest concepts.
- People who play games more than 20 hours per week out-performed people who play games between 5 to 20 hours per week and those who play games less than 5 hours a week by 26% and 65%, respectively; there was no increment in their memory capabilities (based on their self-reporting memory abilities and memory test results).

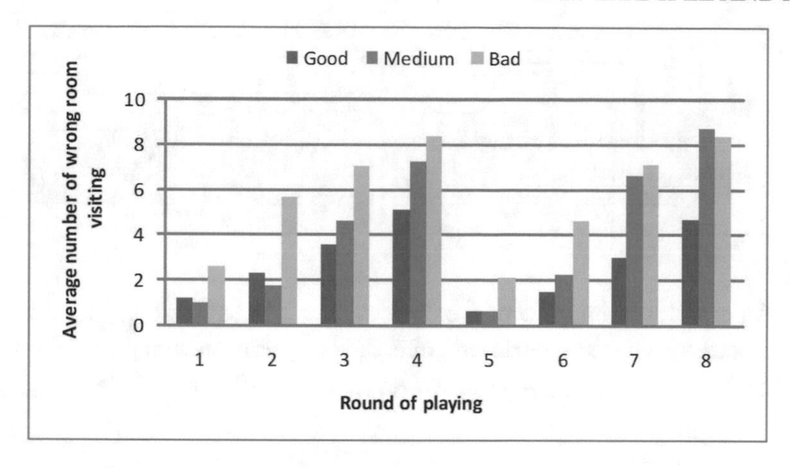

Figure 17.3: Performance results of the user study: subjects are grouped as having good, medium, or bad memories according to the memory test. The first four (i.e., 1 to 4) rounds contain limited objects with different colors while the other four (i.e., 5 to 8) contain more object models in which influence of the color factor was dropped.

To further provide users some guidance in tuning the parameters of the memory model and also to evaluate the sensitivity of these parameters, we conducted several analyses. In the first, we examine the impact of changing values of $Node_{threshold}$ and $Edge_{threshold}$, on how many nodes and edges in their strong stages will get retrieved from the long-term memory into the working memory. These values play a vital role in determining agent memory capability. To carry out the experiment, we chose 100 nodes and 2,500 random edges with randomly assigned strength factors residing in the long-term memory. The first result is shown in Fig. 17.4(a). In this case, the $Edge_{threshold}$ and $SM_{capacity}$ have been set to 5 when $Node_{threshold}$ increases from 1 to 10. The value of $Node_{threshold}$ grows which indicates that the range of node strong stage shrinks, and the number of retrieved strong nodes and strong edges decreases. The result in Fig. 17.4(b) enlarges the difference between the two values; in this setting, both the $Node_{threshold}$ and $SM_{capacity}$ have been set up to 5, while the $Edge_{threshold}$ increases from 1 to 10. The results show enhanced memory capability of the virtual characters: $Node_{threshold}$ has more influence over $Edge_{threshold}$ even though the later value decides how many possible links can be stretching out from a particular node during the memory retrieval process.

Based on the results of our first analysis, the second one tested the total number of retrieved strong nodes and edges in the working memory when the edges were increased from 500 to 2,500. The result is shown in Fig. 17.5. In this case, the $Node_{threshold}$, $Edge_{threshold}$ and $SM_{capacity}$ have been all set to 5 for a consistent experiment. When 500 edges exist in the graph and also with random strength factors that could spread with value equal or higher than 5, we can see basically only starting nodes and nearly no edges would be activated. After that, the difference between the two values increases.

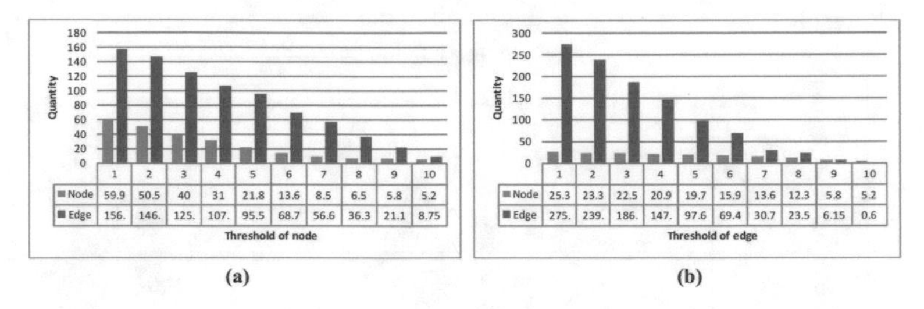

Figure 17.4: An experiment on influences of $Node_{threshold}$ and $Edge_{threshold}$ on the total number of strong nodes and strong edges retrieved from the long-term memory into the working memory. (100 nodes, 2,500 edges; $SM_{capacity} = 5$, $Edge_{threshold}$ in (a) and $Node_{threshold}$ in (b) are both 5).

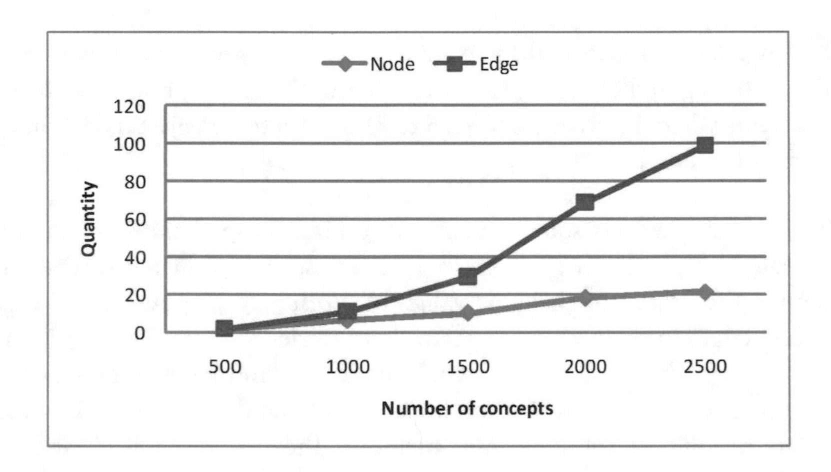

Figure 17.5: An experiment on correlation between number of concepts and total number of nodes and edges in working memory ($SM_{capacity} = 5$, $Node = Edge_{threshold} = 5$).

The last analysis concerns $SM_{capacity}$. While the number of starting nodes controlled by $SM_{capacity}$ is assumed to have an impact on the number of strong nodes and strong edges retrieved into the working memory, we found that increasing the value of $SM_{capacity}$ from 5 to 9 has a trivial effect. This is because while $SM_{capacity}$ is increasing from 5 to 9, making the starting nodes in the activation process increase, this ratio compared to overall nodes (i.e., 100 in this experiment) is still quite small. In addition, the sparse nature of the graph limits the effects.

17.3 FUTURE WORK

We designed and developed a synthetic, parameterizable memory model with the intent of creating more plausibly human agents for games and simulations. Our model embodies a sensory system, working memory model, and long-term memory model, which supports known psychological activities such as forgetting concepts and creating false memories. This allows our virtual characters to perform complex, emergent, and believable behaviors, and promotes human-like variation across a crowd of agents.

This model is incomplete. We could expand the perception model to include modalities other than vision. Memory distortions could be expanded beyond a simple probability model. This mechanism can potentially stimulate agent creativity and enable more spontaneous, emergent behaviors. Given our current detailed memory infrastructure, further development of an effective, plausible mental control module is also worth exploration. Finally, with some optimization, integration with an existing knowledge base could yield interesting results.

CHAPTER 18
Individual Differences

Funda Durupinar, Nuria Pelechano, Jan M. Allbeck, Weizi Li, and Norman I. Badler

18.1 PERSONALITY

Personality is the sum of a person's behavioral, temperamental, emotional, and mental traits. A popular formalization of personality is the Five Factor, or OCEAN (openness, conscientiousness, extroversion, agreeableness, and neuroticism) model. The personality space is composed of these five orthogonal dimensions.

- *Openness* concerns the imaginative and creative aspect of human character. Appreciation of art, inclination towards going through new experiences, and curiosity are characteristics of an open individual.
- *Conscientiousness* determines the extent to which an individual is organized, tidy, and careful.
- *Extroversion* is related to the social aspect of human character.
- *Agreeableness* is a measure of friendliness, generosity, and the tendency to get along with other people.
- *Neuroticism* refers to emotional instability and the tendency to experience negative emotions. Neurotic people tend to be too sensitive and they are prone to mood swings.

Each factor is bipolar and composed of several traits, which are essentially the adjectives that are used to describe people [76]. Some of the relevant adjectives describing each of the personality factors for each pole are given in Table 18.1.

Incorporating the OCEAN personality model into a high-density crowd simulation creates plausible variations in the crowd and enables intuitive user control over these variations. The crowd may consist of subgroups with different collective personalities. Variations in the characteristics of subgroups influence emergent crowd behavior. The user can add any number of groups with shared personality traits and can edit these characteristics during the course of an animation.

Our implementation maps these trait terms to the low-level behavior parameters in the HiDAC (High-Density Autonomous Crowds) crowd simulation system [208]. HiDAC models individual differences by assigning each person psychological and physiological traits. Users normally set these parameters to model the non-uniformity and diversity of a crowd. Our approach

Table 18.1: Trait-descriptive adjectives

O+	Curious, alert, informed, perceptive
O-	Simple, narrow, ignorant
C+	Persistent, orderly, predictable, dependable, prompt
C-	Messy, careless, rude, changeable
E+	Social, active, assertive, dominant, energetic
E-	Distant, unsocial, lethargic, vigorless, shy
A+	Cooperative, tolerant, patient, kind
A-	Bossy, negative, contrary, stubborn, harsh
N+	Oversensitive, fearful, dependent, submissive, unconfident
N-	Calm, independent, confident

frees users of the tedious task of low-level parameter tuning by combining all these behaviors in distinct personality factors.

In order to verify the plausibility of our OCEAN parameter mapping we evaluated users' perception of the personality traits in the generated animations. We created several animations to examine how modifying the personality parameters of subgroups affects global crowd behavior. We found a high correlation between parameters and the user's perception of these trait terms in the videos.

18.1.1 PERSONALITY-TO-BEHAVIOR MAPPING

We now describe the crowd personality mapping in detail. An agent's personality π is a five-dimensional vector, where each dimension is represented by a personality factor, ψ_i. The distribution of the personality factors in a group of individuals is modeled by a Gaussian distribution function N with mean μ_i and standard deviation σ_i:

$$\pi = < \psi_O, \psi_C, \psi_E, \psi_A, \psi_N > \quad (18.1)$$

$$\psi_i = N(\mu_i, \sigma_i^2), \ for\ i \in \{O, C, E, A, N\}, \quad (18.2)$$

where $\mu \in [0, 1]$ and $\sigma \in [-0.1, 0.1]$.

An individual's overall behavior β is a combination of different behaviors. Each behavior is a function of personality as:

$$\beta = (\beta_1, \beta_2, \ldots, \beta_n) \quad (18.3)$$

$$\beta_j = f(n), \ for\ j = 1, \ldots, n \quad (18.4)$$

Since each factor is bipolar, ψ can take both positive and negative values. For instance, a value of 1 for extroversion means that the individual has extroverted character; whereas a value of -1 means that the individual is highly introverted.

By analyzing the meaning and usage of each low-level parameter and built-in behavior in the HiDAC model, we characterize these by the adjectives that are used to describe personalities. This results in a mapping between the agents' personality factors (adjectives) and the HiDAC parameters, as shown in Table 18.2. A positive factor takes values in the range $[0.5, 1]$, while a negative factor takes values in the range $[0, 0.5)$. An unsigned factor indicates that both poles apply to that behavior. For instance E+ for a behavior means that only extroversion is related to that behavior; introversion is not applicable. As indicated in Table 18.2, a behavior can be defined by more than one personality dimension. The more adjectives of a certain factor are defined for a behavior, the stronger the impact of that factor on that behavior. We assign a weight to the factor's impact on a specific behavior. The sum of the weights for a specific type of behavior is 1. We give four representative mappings from a personality dimension to a specific type of behavior as illustrations; the others are mathematically similar and can be found in [51].

Table 18.2: Low-level parameters vs. trait-descriptive adjectives

Leadership	Dominant, assertive, bossy, dependable, confident, unconfident, submissive, dependent, social, unsocial	E, A-, C+, N
Trained/not trained	Informed, ignorant	O
Communication	Social, unsocial	E
Panic	Oversensitive, fearful, calm, orderly, predictable	N, C+
Impatience	Rude, assertive, patient, stubborn, tolerant, orderly	E+, C, A
Pushing	Rude, kind, harsh, assertive, shy	A, E
Right preference	Cooperative, predictable, negative, contrary, changeable	A, C
Avoidance /personal space	Social, distant	E
Waiting radius	Tolerant, patient, negative	A
Waiting timer	Kind, patient, negative	A
Exploring environment	Curious, narrow	O
Walking speed	Energetic, lethargic, vigorless	E
Gesturing	Social, unsocial, shy, energetic, lethargic	E

Right preference. When the crowd is dispersed, individuals tend to look for avoidance from far away and they prefer to move towards the right hand side of the obstacle they are about to face. This behavior shows the individual's level of conformity to the rules: a disagreeable or non-conscientious agent makes a right or left preference with equal probability, while the probability of choosing the right side increases with increase in values of agreeableness and conscientiousness. Given $P_i(Right) \propto A, C$ and $\beta_i^{Right} \in \{0, 1\}$, right preference $P_i(Right)$ is computed as follows:

$$P_i(Right) = \begin{cases} 0.5 & \text{if } \psi_i^A < 0 \; or \; \psi_i^C < 0 \\ \omega_{AR}\psi_i^A + \omega_{CR}\psi_i^C & \text{otherwise} \end{cases} \quad (18.5)$$

$$\beta_i^{Right} = \begin{cases} 1 & \text{if } P_i(Right) \geq 0.5 \\ 0 & \text{otherwise} \end{cases} \quad (18.6)$$

Personal space. Personal space determines the territory in which an individual feels comfortable. Agents try to preserve their personal space when they approach other agents and when other agents approach from behind. However, these two values are not the same. According to the research on Western cultures, the average personal space of an individual is found to be 0.7 meters in front and 0.4 meters behind [84]. Given $\beta_i^{PersonalSpace} \propto^{-1} E$ and $\beta_i^{PersonalSpace} \in \{0.5, 0.7, 0.8\}$, the personal space of an agent i with respect to an agent j is computed as follows:

$$\beta_{i,j}^{PersonalSpace} = \begin{cases} 0.8 \; f(i,j) & \text{if } \psi_i^E \in [0, \frac{1}{3}) \\ 0.7 \; f(i,j) & \text{if } \psi_i^E \in [\frac{1}{3}, \frac{2}{3}] \\ 0.5 \; f(i,j) & \text{if } \psi_i^E \in (\frac{2}{3}, 1] \end{cases} \quad (18.7)$$

$$f(i,j) = \begin{cases} 1 & \text{if } i \; is \; behind \; j \\ \frac{0.4}{0.7} & \text{otherwise} \end{cases} \quad (18.8)$$

Waiting radius. In an organized situation, individuals tend to wait for space available before moving. This waiting space is called the waiting radius and it depends on the kindness and consideration of an individual, i.e., the agreeableness dimension. Given $\beta_i^{WaitingRadius} \propto A$ and $\beta_i^{WaitingRadius} \in \{0.25, 0.45, 0.65\}$, the waiting radius is computed as follows:

$$\beta_{i,j}^{WaitingRadius} = \begin{cases} 0.25 & \text{if } \psi_i^A \in [0, \frac{1}{3}) \\ 0.45 & \text{if } \psi_i^A \in [\frac{1}{3}, \frac{2}{3}] \\ 0.65 & \text{if } \psi_i^A \in (\frac{2}{3}, 1] \end{cases} \quad (18.9)$$

Walking speed. The maximum walking speed is determined by an individual's energy level. Extroverts tend to be more energetic while introverts are more lethargic, so this parameter is controlled by the extroversion trait. Given $\beta_i^{WalkingSpeed} \propto E$ and $\beta_i^{WalkingSpeed} \in [1, 2]$, the walking speed is computed as follows:

$$\beta_i^{WalkingSpeed} = \psi_i^E + 1, \quad (18.10)$$

18.1.2 USER STUDIES ON PERSONALITY

User studies were conducted to evaluate the perceived validity of the suggested mappings. We created several animations to see how global crowd behavior is affected by modifying the personality parameters of subgroups.

Experiment Design

We created 15 videos presenting the emergent behaviors of people in various scenarios where the crowd behavior is driven by the settings of the OCEAN model. We performed the mapping from HiDAC parameters to OCEAN factors by using trait-descriptive adjectives. We determined the correspondence between our mapping and user perceptions of these trait terms in the videos. Seventy subjects (21 female, 49 male, ages 18–30) participated in the experiment. We showed the videos to the participants through a projected display and asked them to fill out a questionnaire consisting of 123 questions—about eight questions per video. The videos were shown one by one; after each video, participants were given some time to answer the questions related to the video. The participants did not have any prior knowledge about the experiment. Questions assessed how much a person agreed with statements such as "I think the people in this video are kind." or "I think the people with black suits are calm." We asked questions that included the adjectives describing each OCEAN factor instead of asking directly about the factors because we assumed that the general public might be unfamiliar with the OCEAN model. Participants chose answers on a scale from 0 to 10, where 0 = totally disagree, 5 = neither agree nor disagree, and 10 = totally agree. We omitted the antonyms from the list of adjectives for the sake of conciseness. The remaining adjectives used were *assertive*, *calm*, *changeable*, *contrary*, *cooperative*, *curious*, *distant*, *energetic*, *harsh*, *ignorant*, *kind*, *orderly*, *patient*, *predictable*, *rude*, *shy*, *social*, *stubborn*, and *tolerant*. The sample scenarios can be seen in Fig. 18.1.

Analysis

After collecting all participant responses for all the videos, we first organized the data for the adjectives. Each adjective is classified by its question number, the actual simulation parameter, and the participant answers for the corresponding question. We calculated the Pearson correlation (r) between the simulation parameters and the average of the subject answers for each question.

We grouped the relevant adjectives for each OCEAN factor to assess the perception of personality traits. The evaluation process is similar to the evaluation of adjectives; this time considering the questions for all the adjectives corresponding to an OCEAN factor. For instance, as openness is related to curiosity and ignorance, we took into account the adjectives *curious* and *ignorant*. Again, we averaged the subject answers for each question. We computed the correlation with the parameters and the mean throughout all the questions inquiring *curious* and *ignorant*.

We computed the significance of the correlation coefficients as $1 - p$, where p is the two-tailed probability that is calculated considering the sample size and the correlation value. Higher correlation and significance values suggest more accurate user perceptions.

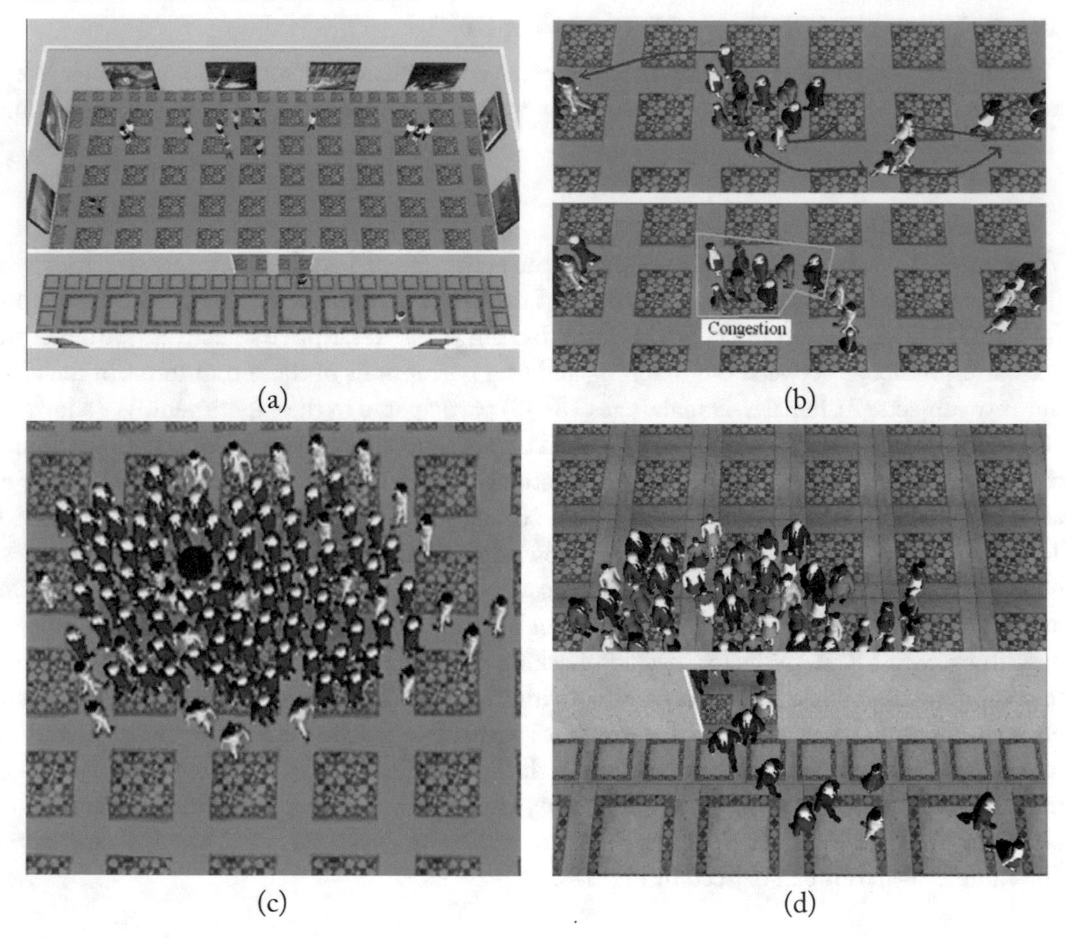

Figure 18.1: Snapshots of a crowd simulation authored using our framework: (a) Openness tested in a museum. The most open people (red-heads) stay the longest, whereas the least open people (blue-heads) leave the earliest. (b) People with low conscientiousness and agreeableness values cause congestion.(c) Ring formation where extroverts (blue suits) are inside and introverts are outside. (d) Neurotic, non-conscientious and disagreeable agents (in black suits) show panic behavior.

Results

Figure 18.2(a) depicts the correlation coefficients and significance values for the adjectives. Significance is low (< 0.95) for *changeable*, *orderly*, *ignorant*, *predictable*, *social*, and *cooperative*. Low significance is caused by low correlation values for *changeable* and *orderly*. However, although the correlation coefficients are found to be high for *predictable*, *ignorant*, *social* and *cooperative*, low significance can be explained due to small sample size. From the participant comments, we determined that *changeable* is especially confusing because the participants identified non-conscientious agents as rude but perceived them as persistent in their rudeness.

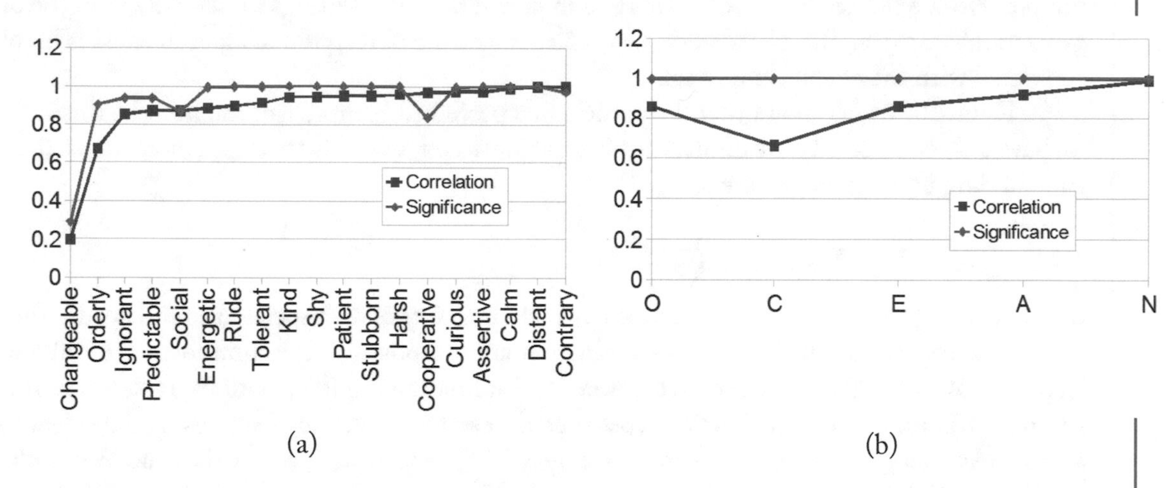

(a)

(b)

Figure 18.2: (a) The correlation coefficients between the parameters and the subject answers for the descriptive adjectives (blue), and the significance values for the corresponding correlation coefficients (orange). Significance is low (< 0.95) for changeable, orderly, ignorant, predictable, social, and cooperative. (b) The correlation coefficients between actual parameters and subject answers for the OCEAN factors (blue), and the two-tailed probability values for the corresponding correlation coefficients (orange). All the coefficients have high significance.

Orderly is another weakly correlated adjective. Analyzing the results for each video, we found that agents in the evacuation scenario were perceived to be orderly although they displayed panic behavior. In these videos, even if the agents pushed each other and moved fast, some kind of order could be observed. This was due to the smooth flow of the crowd during building evacuation. Although people were impatient and rude, the overall crowd behavior appeared orderly. On the other hand, in a scenario showing queuing behavior in front of a water dispenser, the participants could easily distinguish orderly agents from disorderly ones. Orderly agents waited at the end of the queue, while disorderly agents rushed to the front. In this scenario, although the main goal was the same for all the agents (drinking water), there were two distinguishable groups that acted differently.

Figure 18.2 (b) shows the correlation coefficients and their significance for the OCEAN parameters. These values are computed by taking into account all the relevant adjectives for each OCEAN factor. All the coefficients have high significance, with a probability of less than 0.5% of occurring by chance ($p < 0.005$). The significance is high because all the adjectives describing a personality factor are taken into account, achieving sufficiently large sample size.

The correlation coefficient for conscientiousness is comparatively low, showing that the participants correctly perceived only approximately 44% of the traits ($r^2 \approx 0.44$). Low correlation values for *orderly* and *changeable* reduce the overall correlation. If we consider only *rude* and *predictable* for conscientiousness, correlation increases by 18.6%. The results suggest that people can observe the politeness aspect in short-term crowd behavior settings more easily than the organizational aspects. This observation also explains why the perception of agreeableness is highly correlated with the actual parameters.

Figure 18.2 also shows that the participants perceived neuroticism the best. In this study, we have only considered the calmness aspect of neuroticism, which is tested in emergency settings and building evacuation scenarios.

18.2 ROLES AND NEEDS

Crowd heterogeneity can also be addressed by including agents with social roles and needs. These characteristics also promote agents with function and purpose. This section focuses on higher-level control mechanisms as opposed to lower-level animation implementations. It also links roles and role switching to different action types such as reactions, scheduled actions, and need-based actions. As such the authoring of roles is largely just associating a set of these actions with a role. The techniques and methodologies used are adopted from a number of research disciplines including multi-agent systems, social psychology, ontologies, and knowledge representations, as well as computer animation.

18.2.1 APPROACH

Here we provide an overview of our approach and describe the social psychological models on which it is founded, including a definition of roles, factors affecting role switching, and action types.

Definition of Role

A role is *the rights, obligations, and expected behavior patterns associated with a particular social status* [45]. Ellenson's work [56] notes that each person plays a number of roles. Taking into consideration these descriptions as well as discussions from other social psychologists [18, 183], we conclude that roles are patterns of behaviors for given situations or circumstances. Roles can demand certain physical, intellectual, or knowledge prerequisites, and many roles are associated with social relationships.

Role Switching

A person's priorities are set by a number of interacting factors, including emotions, mood, personality, culture, roles, status, needs, perceptions, goals, relationships, gender, intelligence, and history, just to name a few. Here we have chosen only a few factors that are related to roles and role switching that we believe will help endow virtual humans with meaningful, purposeful behaviors.

Switching from one role to another can be linked to time, location, relationships, mental status, and needs (Fig. 18.3). For example, one can imagine someone switching to a *businessman* role as the start of the work day approaches or as he enters his office or when he encounters his boss. Also, someone may need to shop for groceries to provide for his family. The *shopping* behavior would stem from a need and cause a switch in role to *shopper*. Elements of mental status, such as personality traits, can impact the selection and performance of these roles. For example, a non-conscientious person might not shop for groceries even if the need exists.

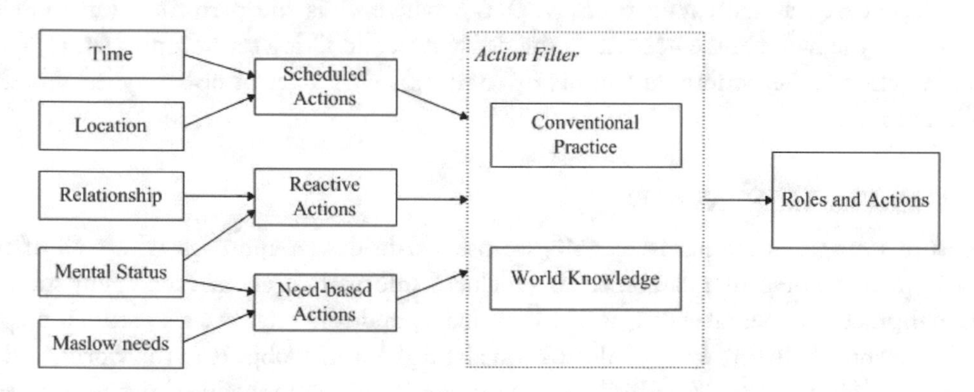

Figure 18.3: System diagram.

Action and role selection is further affected by a filtration of proposed actions according to an agent's *Conventional Practice* and *World Knowledge* [18, 101]. *Conventional Practice* is a set of regulations and norms that each individual in the society should obey. *World Knowledge* indicates that certain physical, intellectual, and knowledge elements are required for specific roles.

Another perspective on action selection is to examine what triggers various behaviors. Some actions are planned, such as going to work or attending a meeting. These actions, called *Scheduled* actions, tend to establish a person's daily routine and are often heavily coupled with their roles. Other actions are not so predetermined.

Some actions arise to fulfill needs. Among these needs might be those depicted in *Maslow's Hierarchy of Needs*, including, food, water, excretion, friendship, family, employment, and creativity [176]. This action type, called *Need-based*, can also be linked to roles. For example, in order to maintain employment safety, a businessman might need to contact his clients on a regular basis. Still other actions, *Reactive actions*, are responses to agents, objects, or events in the world. Who and what we react to is at least partially determined by our roles. If we see a friend or co-worker as we are walking to work, we are likely to stop and greet them.

While we cannot claim that these three types of actions make a complete categorization of all behaviors, we believe that they can encompass a wide range of behaviors and provide strong ties to roles and purposeful behaviors. Another key factor is the ability to easily author or initiate these behaviors. Each requires a finite, straightforward amount of data:

Scheduled actions: $Sch = \langle P, A, L, T \rangle$, where P is the performer (an individual or group), A is the action to be performed (simple or complex), L is the location where the action is to be performed (based on an object or a location), and T is an indication of the time (i.e., start time and duration).

Reactive actions: $Rea = \langle P, S, A \rangle$, where P is the performer (an individual or group), S is the stimulus (an object, type of object, person, location, event, etc.) and A is the action to be performed (simple or complex).

Need-based actions: $Nee = \langle P, N, D, C \rangle$, where P is the performer (an individual or group), N is the name of the need, D is the decay rate, and C is a set of tuples $\langle A, O, F \rangle$, where A is the action to be performed (simple or complex), O is a set of object types, and F is the fulfillment rate.

18.2.2 IMPLEMENTATION

To ensure scenario authoring is feasible, we use a data-driven approach where all of the vital scenario data is stored in a database. This includes information about each agent such as conventional practices, mental status, world knowledge, and role sets. We also store a mapping of the relationships between agents. Information about the world, objects in the world, and actions are also stored in the database. This includes the specification of schedules, needs, and reactions. As such, scenarios can be authored entirely through the database and without any coding. These extensions are built on an existing crowd simulator [207].

We also enriched our previous definition of role. The most important component of a role is a set of actions [183]. This action set corresponds to the conventional practices associated with the role. These actions may be scheduled, need-based, or reactive actions. Furthermore, these actions may also be linked to parameters such as location, object participants, start-times, and durations. Interpersonal roles are also associated with relationships, which are a simple named linking of agents. Among agent parameters is a set of capabilities corresponding to actions that they can perform. These capabilities form the foundation of an agent's world knowledge.

Factors influencing role selection include time, location, relationships, mental status, and needs. In this section, we will describe the implementation of each of these factors and how they have been incorporated into the three action types. We will also discuss how actions and roles are further filtered by conventional practice and an agent's world knowledge (Fig. 18.3).

Action Types

A large part of the definition of a role includes a pattern of behaviors. Our framework associates each role with a set of actions from the sets *Scheduled*, *Reactive*, or *Need-based*.

Scheduled actions include time and location parameters [3] and can be used to establish an agenda for a day. Some roles are directly associated with scheduled actions. For example, a businessman may be scheduled to work in his office from 9am to 5pm. As 9am approaches, the framework will initiate processing of the scheduled work in office action and send the character to his office and the businessman role. However, if an agent does not have a scheduled action to perform, it will perform a default action that is associated with their current role. Generally default actions are the actions most often performed by that role. For example, a businessman or administrator might work in an office. A shopkeeper might attend to the cash register. Just as in real life, scheduled actions can be suspended by higher priority need-based and reactive actions.

Need-based actions are merely database entries associating a decay rate, actions, objects, and a fulfillment quotient. Conceptually, there is a reservoir that corresponds to each need for each agent. The initial level of each reservoir is set randomly at the beginning of the simulation. At regular intervals, the reservoirs are decreased by the specified decay rates. When the level of a reservoir hits a predetermined threshold, the fulfilling action is added to the agent's queue of actions. Its priority will increase as the reservoir continues to decrease; eventually its priority will be greater than all actions on the queue. Then the agent will perform the action, raising the level of the reservoir. Another example of a reservoir-based system is described in [249].

We have chosen to use *Mental Status* as an influence on needs (and reactions), because social scientists have linked it to roles and we feel it adds plausible variability. Mental status includes several factors, but we focus on personality as it addresses an individual's long-term behavior. Again, we use the OCEAN model. Needs and priorities differ from person to person. We represent this variation by linking the personality traits with needs. This is, of course, a massive oversimplification, but one that leads to plausible variations. Table 18.3 shows our mapping from Maslow needs to OCEAN personality dimensions. Note that just the personality dimension is represented, not the valence of the dimension. For example, neuroticism is negatively correlated with needs for security of employment and family. This mapping was formulated by examining the adjectives associated with the personality dimensions and the descriptions of the Maslow needs. More details about the mapping of needs to personality traits can be found in [163].

Reactive actions: As a simulation progresses, agents make their way through the virtual world, attempting to adhere to their schedules and meet their needs. In doing so, agents encounter many stimuli to which a reaction might be warranted. Reactions play an important part in our implementation of roles. In [59], Merton states that a person might switch roles as a response to those around him. Relationships are a major impetus for reactive role switching. For example, if two agents encounter each other and are linked by a relationship such as friendship, they will switch to the friend role.

Action Filter

Once a set of actions and roles have been proposed, some may be eliminated due to conventional practice constraints or an agent's lack of necessary world knowledge.

Conventional Practice Social science researchers believe that when an agent plays a role in a given organizational (or social) setting, she must obey *Conventional Practices*, meaning *behavioral constraints* and *social conventions* (e.g., the businessperson must obey the regulations that the company stipulates) [186, 192]. To be more specific, behavioral constraints are associated with the following factors: responsibilities, rights, duties, prohibitions, and possibilities [192]. Role hierarchies include conventional roles (e.g., *citizen, businessperson, sales clerk*) and interpersonal roles

Table 18.3: Mapping between Maslow need reservoirs and personality dimensions

Reservoir Descriptions	Personality Traits
problem solving, creativity, lack of prejudice	O, A
achievement, respect for others	O, A
friendship, family	E
security of employment, security of family	C, N
water, food, excretion	

(e.g., *friends, lovers, enemies*). We have linked each conventional practice norm with an impact factor (range [0, 1]) which reflects how strongly these norms are imposed on certain roles.

World Knowledge Some roles have physical or intellectual requirements and that may be difficult to ascertain. Some people are also just naturally more physically or intellectually gifted or have more talent in an area than others. These factors can put limitations on what roles a person can take on [18]. We represent world knowledge as capabilities. Agent capabilities are the set of actions that the agent can perform. Actions are categorized and placed in a hierarchy to lessen the work of assigning capabilities to agents and also checking to ensure that agents meet the capability conditions before performing an action.

Examples

To explore the effects of roles and more precisely role switching on virtual human behaviors, we have authored a typical day in a neighborhood. As with real people, each virtual human is assigned a set of roles. Figure 18.4 demonstrates one agent taking on the role of *businessman* as he enters his office building (i.e., location-based role switching). Two other agents react to seeing each other by switching to *friend* roles (i.e., relationship-based role switching). Another agent reacts to trash in the street by starting her *cleaner* role (i.e., behavior selection-based role switching).

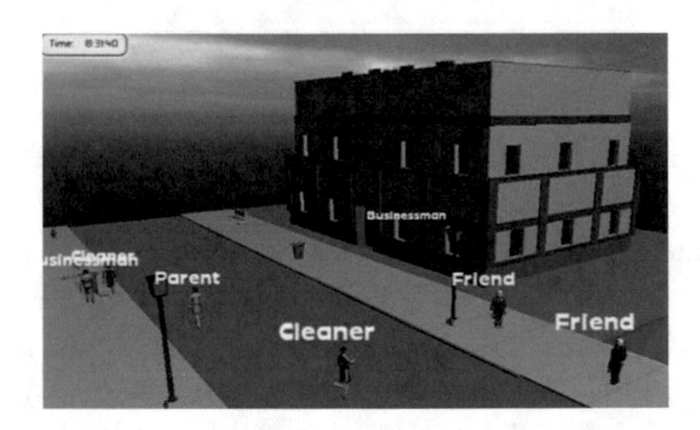

Figure 18.4: Locations, relationships, and behavior selection affect role switching.

These examples of agents going to work demonstrate how time, location, and behavior selection impact role switching. They focus on role transitions caused by scheduled and reactive actions. The following office examples concentrate more on need-based actions. The top image of Fig. 18.5a shows that the businessman's *creativity* reservoir is approaching the critical threshold (i.e., 2). When it reaches the threshold, he suspends his current action and starts a conversation with his co-worker. This exchange of ideas causes the *creativity* reservoirs of both men to refill.

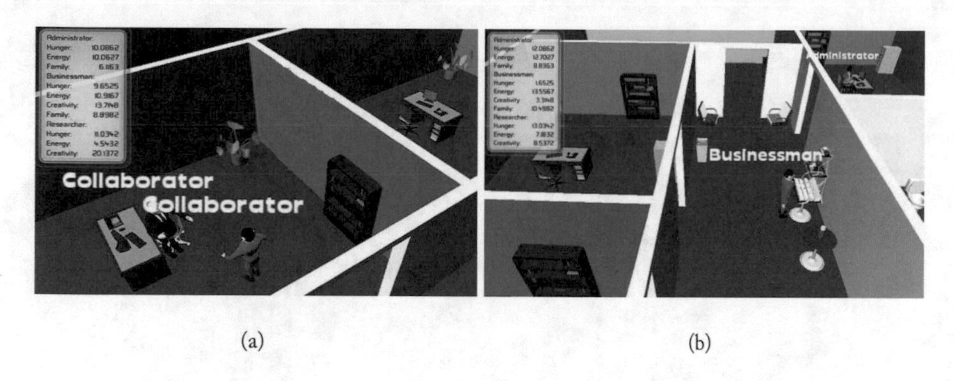

(a) (b)

Figure 18.5: (a) The businessman's *creativity* need prompts him to speak to a co-worker, causing both to switch to *collaborator* roles and replenish their creativity reservoirs. (b) The businessman's role remains while eating to refill his hunger reservoir.

Figure 18.5b shows that not all need-based actions cause role switching. The hunger need is associated with the role of being human. Because *businessman* is a descendant of this role in the role hierarchy, there is no need to switch.

CHAPTER 19

Conclusion

Crowd simulations can better depict real populations if they include agent traits that introduce character heterogeneity. The memory model presented is one avenue for differentiating individuals. Agents are capable of remembering objects and their locations they encounter in the environment. These memories can be strengthened through repeated encounters and can be forgotten or mis-remembered based on models of how human memory works. Agent memories can impact navigation [208], overall behavior selection, and decision-making. Future work should include examining the impact of various memory model parameter values on the resulting crowd simulation *realism*.

The OCEAN personality model defines five orthogonal axes to intuitively describe human personalities. We described a mapping of these personality traits to low-level simulation parameters, facilitating the control of agent and group personality in order to observe their consequences on crowd behaviors. We validated our mapping by conducting a user study which assesses the perception of personality traits in a variety of crowd simulations which demonstrate these behaviors. The personality traits provide an intuitive and flexible interface for authors to control social crowd dynamics. These same personality traits can also be used to vary behavior selection in a functional crowd simulation and influence other psychological models like the modeling of needs.

We presented a framework for instilling virtual humans with roles and role switching, with the intention of having agents portray more purposeful behaviors. The roles are based on social psychology models and focus on approaches that facilitate authoring flexibility. As real life complexities rarely allow agents to embody just a single role during the course of a day, role switching is important to creating behavioral variety and richness. The framework presented can also be used to include abnormal behaviors. For example, one could author a subversive role for a character that includes reacting to pedestrians by robbing them or includes a strong need for drugs and alcohol. There are numerous possible extensions to this work. First, we could illustrate the dynamics that stem from status hierarchies by experimenting with the concepts of power scale and social distance. We might also address situations where multiple roles could be adopted. For example, a man being approached by both his boss and his child. Here, social power scales and social distance might result in different social threats which would cause one role to be favored over the other. Finally, we could focus on agents that learn role definitions by observing the behaviors of others, enabling each agent to have customized definitions based on their own experiences.

We have described just a few potential psychological models that could be used to create more heterogeneous crowds. Others may have just as much impact, including emotions [9]. The crowd simulation application domain should dictate what models should be included and which are not required. Future studies might explore the space of available psychological models and crowd simulation applications to provide insight on optimally beneficial combinations.

PART V

Behavior and Narrative

CHAPTER 20

Background

There exists a wealth of research [118, 208, 272] that separately addresses the problems of character animation, steering, navigation, and behavior authoring, with many open challenges [125] that need to be addressed in an effort to arrive at a common framework for simulating autonomous virtual humans for the next generation of narrative-driven interactive virtual world applications.

The next generation of interactive virtual worlds demand functional, purposeful characters with diverse personalities [126]. Crowd approaches [208, 272] provide interfaces for generating ambient background activity but are not suitable for authoring principal actors that drive the narrative forward. The work of Kwon et al. [149], and Kim et al. [138, 140] synthesizes synchronized multi-character motions and crowd animations by editing and stitching motion capture clips. "Motion patches" [158] annotate motion semantics in environments and can be concatenated [139] or precomputed [247] to synthesize complex multi-character interactions. Recent work [301] provides sophisticated tools for generating and ranking complex interactions (e.g., fighting motions) between a small number of characters from a text-based specification of the scene. The focus of these approaches is to produce locally rich, complex, and populated scenes [106] where the consistency of the interactions between characters toward an overarching narrative is less relevant.

Animating behaviors in virtual agents has been addressed using multiple diverse approaches, particularly with respect to how behaviors are designed and animated. Early work focused on imbuing characters with distinct, recognizable personalities using goals and priorities [173] along with scripted actions [212].

The problem of managing a character's behavior can be represented with decision networks [306], cognitive models [65], goal-oriented action planning [131, 132, 304], or learning [165]. Very simple agents can also be simulated on a massive scale using GPU processing [58]. Recent work [242, 261] uses an event-centric authoring paradigm to facilitate complex multi-actor interactions, but isolated events by themselves do not comprise a connected story arc.

Scripted approaches [173, 179] describe behaviors as pre-defined sequences of actions. However, small changes in scripting systems often require extensive modifications of monolithic specifications. Systems such as Improv [212], LIVE [185], and "Massive" use rules to define an actor's response to external stimuli. These systems require domain expertise, making them essentially inaccessible to end-users, and are not designed for authoring complicated multi-actor interactions over the course of a lengthy narrative.

Interactive narrative is a rich research topic with numerous and diverse methodologies. One of the earliest planning systems, the STRIPS planner [64], lays down the framework of a described world state with operators, preconditions, and effects manipulating that state. Subsequent developments to the pervasive STRIPS archetype include the planning domain definition language [180], cognitive-oriented planning [68], hierarchical task networks [57], and planning with smart objects [1]. Kapadia et al. [130, 131] uses domain-independent planning for multi-actor behavior authoring.

Total-order planners [64, 85] are promising for automated behavior generation. These approaches require the specification of domain knowledge, and sacrifice some authoring precision, but they permit the automatic generation of a strict sequence of actions to satisfy a desired narrative outcome. Planning in the action space of all participating actors scales combinatorially, and these approaches are restricted to simplified problem domains with small numbers of agents [104]. Partial-order planners [233] relax the strict ordering constraints during planning to potentially handle more complex domains, and have been applied to accommodate player agency in interactive narratives [31, 121].

Interactive narrative systems draw on these techniques and others to create virtual worlds with a narrative focus. Façade [179], the first fully realized interactive narrative system, uses natural language understanding and pre-authored narrative beats to create a story that adheres to an Aristotelian narrative arc. Mimesis [228] uses narrative planning [161] with atomic agent actions, Thespian [248] uses decision-theoretic agents, and PaSSAGE [275] guides a player through prescripted "encounters" based on the system's estimation of the player's ideal experience. Most of these systems employ some form of drama manager or director [175], a virtual agent responsible for monitoring the state of the story and intervening according to narrative goals. Riedl and Bulitko [229] also provide a more detailed survey of the current state of the art in interactive narrative.

As we diversify the actions that virtual characters perform, we encounter a combinatorial explosion of those characters' potential interactions. If they are modeled as monolithic autonomous agents, coordinating their activities becomes a challenge of passing messages and sharing state between thousands of virtual peers, especially in prolonged interactions where actors take on roles. In such a system, introducing a new factor to the environment requires updating each character type to account for the newly introduced mechanic. This amplifies the design effort required to introduce new content or author interesting behaviors involving multiple agents.

Scripted approaches to authoring narratives allow for computational efficiency and authorial control at the expense of flexibility and with a great deal of effort in content creation. In contrast, narrative planners that operate in the action space of each agent's individual capabilities quickly suffer from combinatorial explosion with large groups of agents. Additionally, like all highly emergent systems, they may produce unexpected and undesirable narrative plans.

In an ideal interactive virtual world system, we want an untrained content creator to be able to design rich behaviors involving multiple characters interacting with one another and the environment. To facilitate this, we advocate an event-centric approach that can temporarily operate multiple characters as if they were limbs of the same entity. Actors participating in an event suspend autonomy and are exclusively controlled by the event to ensure coordination between actors.

In order to provide a readily available solution for many of the standard practices in virtual human simulation, we developed ADAPT [123, 246], an open-source platform for authoring complex multi-actor behaviors. Chapter 21 describes ADAPT in detail. On top of ADAPT, we utilize an event-centric planner that allows a virtual director to decide which events occur at which locations and with which participants, while still leaving the details of the underlying behavior in the hands of an author. This alleviates the back-and-forth planning complexity of multi-actor interactions, and allows characters to exhibit a rich repertoire of actions (such as gestures, reaching, gazing, responding to external forces, etc., supported through ADAPT) suitable for a complex 3D environment without affecting the branching factor of the simulation. Chapter 22 describes our approach in detail.

CHAPTER 21

An Open Source Platform for Authoring Functional Crowds

Alexander Shoulson, Nathan Markshak, Mubbasir Kapadia, and Norman I. Badler

21.1 ADAPT

Animating interacting virtual humans in real-time is a complex undertaking, requiring the solution to numerous tightly coupled problems such as steering, path-finding, full-body character animation (e.g., locomotion, gaze tracking, and reaching), and behavior authoring. This complexity is greatly amplified as we increase the number and degree of sophistication of characters in the environment. Numerous solutions for character animation, navigation, and behavior design exist, but these solutions are often tailored to specific applications, making integration between systems arduous. Integrating multiple character control architectures requires a deep understanding of each controller's design so that they may communicate with one another; otherwise character controllers will conflict at overlapping parts of the body and produce visual artifacts by naïvely overwriting one another. Directly modifying arbitrary character controllers to cooperate with one another and respond to external behavior commands can be costly and time-consuming. Monolithic, feature-rich character animation systems do not commonly support modular access to a subset of their capabilities, while simpler systems lack control fidelity. Realistically, no sub-task of character control has a "perfect" solution. An ideal character animation system would allow a

Figure 21.1: Demonstrating the capabilities of ADAPT. Visualizing multiple choreographers that blend to produce a pose for the display model; an agent reacting to the impact force of a ball; a crowd of 100 agents resolving a bottleneck; three characters engaged in a conversation.

designer to choose between preferable techniques for producing a particular action or animation, leveraging the wealth of established systems already produced by the character animation research community, and interfacing with robust frameworks for behavior and navigation.

ADAPT is a modular system that allows for the seamless integration of multiple character animation controllers on the same model, without requiring any controller to drastically change or accommodate any other. Rather than requiring a tightly coupled set of character controllers, ADAPT uses a system for blending arbitrary poses in a user-authorable dataflow pipeline. ADAPT couples these animation controllers with an interface for path-finding and steering, as well as a comprehensive behavior authoring structure for both individual decision-making and complex interactions between groups of characters. ADAPT allows the addition of new character controllers and behavior routines with minimal integration effort. Since controllers do not need to be fundamentally redesigned to work with one another, we avoid the combinatorial effect of having to modify each pre-existing controller to adjust for the change. Our system for character control contributes to our core goal of providing a platform for experimentation in character animation, navigation, and behavior authoring. Developers can rapidly iterate on character controller designs with visual feedback, compare their results with other established systems on the same model, and use features from other packages to provide the functionality they lack without the need to deeply integrate or re-invent known techniques. In essence, ADAPT can be the "glue" that integrates all the techniques we have discussed so far.

21.2 FRAMEWORK

ADAPT operates at multiple layers with interchangeable, lightweight components, and we focus on minimizing the amount of communication and interdependency between modules (Fig. 21.2). The animation system performs control tasks such as locomotion, gaze tracking, and reaching as independent modules, called choreographers, that can share parts of the same character's body without explicitly communicating or negotiating with one another. These modules are managed by a coordinator, which acts as a central point of contact for manipulating the virtual character's pose in real-time. The navigation system performs path-finding with predictive steering, but an interface allows users to change the underlying navigation library without affecting the functionality of the rest of the framework. The behavior level is split into two tiers. Individual behaviors are attached to each character and manipulate that character using the behavior interface, while a centralized control structure orchestrates the behavior of multiple interacting characters in real-time. The ultimate product of our system is a pose for each character at an appropriate position in the environment, produced by the animation coordinator and applied to a rendered virtual character in the scene each frame.

21.2.1 FULL-BODY CHARACTER CONTROL

Controlling a fully articulated character is traditionally accomplished using a series of interwoven subcomponents responsible for various parts of the body. Without prior knowledge of other sys-

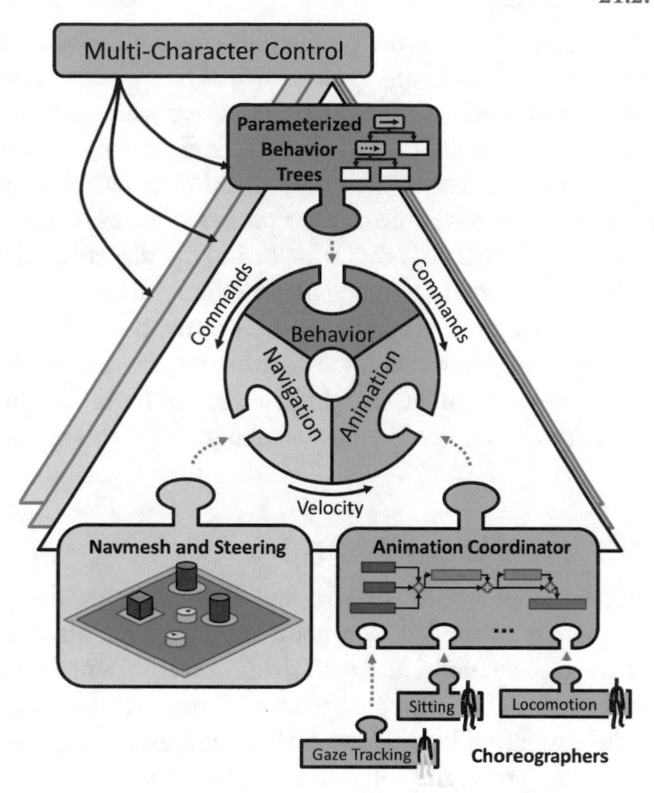

Figure 21.2: Overview of ADAPT, illustrating the structure for controlling an individual character and all of the characters in an environment. Every character has a core interface for behavior, navigation, and animation, each of which connects to more specific modular components. Top-level narrative control communicates with each character through the behavior interface.

tems, a designer creating a character controller will generally do so with thc assumption that no other systems are acting on the rigged model at the same time. If a controller sets the orientation or position of a character's joint, it does so expecting no other controller to overwrite that orientation or position in the current frame. If two controllers conflict and overwrite one another, the constant changes cause visual artifacts such as jitter as the character rapidly shifts between the two settings for its joints. Controllers can be made to share control of a single body either by negotiating with one another, or by dividing the body into sections and controlling those sections independently. However, this requires that the controllers be specifically designed to coordinate, which requires additional effort on either the designer or the user of the control system. The addition of new functionality also becomes more difficult as all of the previous body controllers must be modified to communicate with any new components and share control of the body's joints.

To address this issue, we divide the problem of character animation into a series of isolated, modular components called *choreographers* attached to each character. Each choreographer operates on a *shadow*, which is an invisible clone of the character skeleton, and has unmitigated control to manipulate the skeletal joints of its shadow. Each frame, a choreographer produces an output pose consisting of a snapshot of the position and orientation of each of the joints in its private shadow. A *coordinator* receives the shadow poses from each choreographer and performs a weighted blend to produce a final pose that is applied to the display model for that frame. Since each choreographer has its own model to manipulate without interruption, choreographers do not need to communicate with one another in order to share control of the body or prevent overwriting one another. This allows a single structure, the coordinator, to manage the indirect interactions between choreographers using a simple, straightforward, and highly authorable process centered around blending the shadows produced by each choreographer. This system is discussed in more detail in Section 21.3.

21.2.2 STEERING AND PATH-FINDING

We use a navigation mesh approach for steering and path-finding with dynamic obstacle avoidance. Each display model is controlled by a point-mass system, which sets the root positions (usually the hips) of the display model and each shadow every frame. A navigation interface includes basic commands such as setting a goal position. Character choreographers do not directly communicate with the navigation layer. Instead, choreographers are made aware of the position and velocity of the character's root, and will react to that movement on a frame-by-frame basis. A character's orientation can follow several different rules, such as facing forward while walking, or facing in an arbitrary direction, and we handle this functionality outside of the navigation system itself. ADAPT supports both the Unity3D built-in navigation system and the Recast/Detour library [190] for path-finding and predictive goal-directed collision avoidance, and users can easily experiment with alternate solutions, such as navigation graphs.

21.2.3 BEHAVIOR

ADAPT is designed to accommodate varying degrees of behavior control by providing a diverse set of choreographers and navigation capabilities. Each character has a capability interface with commands like `ReachFor()`, `GoTo()`, and `GazeAt()` that take straightforward parameters like positions in space and send messages to that character's navigation and animation components. To invoke these capabilities, we use Parameterized Behavior Trees (PBTs) [243], which present a method for authoring character behaviors that emphasizes simplicity without sacrificing expressiveness. Having a single, flat interface for a character's action repertoire simplifies the task of behavior authoring, with well-described and defined tasks that a character can perform. One advantage of the PBT formalism is that they accommodate authoring behavior for multiple actors in one centralized structure. For example, a conversation between two characters can be designed in a single data structure that dispatches commands to both characters to take turns playing sounds or gestural animations. For very specific coordination of characters, this approach can be preferable over traditional behavior models where characters are authored in isolation and interactions between characters are designed in terms of stimuli and responses to triggers. ADAPT also generalizes across traditional or experimental new ways of modeling behavior to cover cases where PBTs are not the most appropriate. The behavior system is discussed in more detail in Section 21.4.

21.3 SHADOWS IN FULL-BODY CHARACTER ANIMATION

Modeling systems describe a virtual human as a skinned mesh with a hierarchical skeletal structure underneath. The movement of the body is determined by altering the position and orientation of each joint in the skeleton "rig", which in turn affects the position and orientation of that joint's children in the hierarchy. General character controllers are systems designed to manipulate the character by setting the positions and orientations of that character's joints, either via animations or procedurally with physical models or inverse kinematics. We coordinate these controllers by allocating each choreographer its own private "shadow" character model, a replica of the skeleton (or a subset) of the character being controlled (Fig. 21.3). The animation process has two interleaved steps. First, each choreographer manipulates its personal shadow and outputs a snapshot (called a shadow pose) describing the position and orientation of that shadow's joints at that time step. Then, we use the centralized coordinator to blend the shadow pose snapshots into a final pose for the rendered character. Note that "shadow" refers to the invisible articulated skeleton allocated to each choreographer, while a "shadow pose" is a serialized snapshot containing the joint positions and orientations for a shadow at a particular point in time.

21.3.1 CHOREOGRAPHERS

The shadow pose of a character at time t is given by $\mathbf{P}_t \in \mathbb{R}^{4\times|J|}$, where $\mathbf{P}_t^j$ is the configuration of the jth joint at time t. A choreographer is a function $C(\mathbf{P}_t) \longrightarrow \mathbf{P}_{t+1}$ which produces the next pose by changing the configuration of the shadow joints for that time step. Using these definitions, we define two classes of choreographers:

Generators. Generating choreographers produce their own shadow pose each frame, requiring no external pose data to do so. Each frame, the input shadow pose $\mathbf{P}_t$ for a generator C is the pose $\mathbf{P}_{t-1}$ generated by that same choreographer in the previous frame. For example, a sitting choreographer requires no external input or data from other choreographers in order to play the animations for a character sitting and standing, and so its shadow's pose is left untouched between frames. This is the default configuration for a choreographer.

Transformers. Transforming choreographers expect an input shadow pose, to which they apply an offset each frame. Each frame, the input shadow pose $\mathbf{P}_t$ to a transformer C is an external shadow pose $\mathbf{P}'_{t+1}$ from another choreographer C', computed for that frame. The coordinator sets its shadow's pose to $\mathbf{P}'_{t+1}$ and applies an offset to the given pose during its execution, to produce a new pose $\mathbf{P}_{t+1}$. For example, before executing, the reach choreographer's shadow is

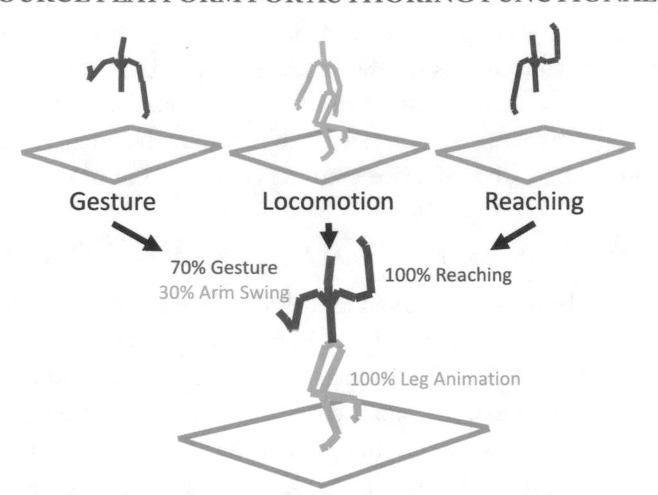

Figure 21.3: Blending multiple character shadows to produce a final output skeleton pose. For example, we combine the pose of the locomotion choreographer (green, full-body) during a walk cycle with the reaching choreographer (red, upper-body) extending the left arm towards a point above the character's head, and the gesture choreographer (blue, upper-body) playing a waving animation. The generated poses are projected, either wholly or partially, on different sections of the displayed body during any particular frame. The partial blend is represented with a mix of colors in the RGB space.

set to the pose of a previously updated choreographer's shadow (say, the locomotion choreographer with swinging arms and torso movement). The reach choreographer then solves the reach position from the base of the arm based on the torso position it was given, and overwrites its shadow's arm and wrist joints to produce a new pose. Without an input shadow, the reach choreographer would not be aware of other choreographers moving the torso, and would not be able to accommodate different torso positions when solving a reaching problem. Note that this is accomplished without the choreographers directly communicating or even being fully aware of one another. A transforming choreographer can receive an input pose, or blend of input poses, from any choreographer that has already been updated in the current frame.

21.3.2 THE COORDINATOR

During runtime, our system produces a pose for the display model each frame, given the character choreographers available. This is a task overseen by the coordinator. The coordinator is responsible for maintaining each choreographer, organizing the sequence in which each choreographer performs its computation at each frame, and reconciling the shadow poses that each choreographer produces by sending them between choreographers and/or blending them together. The coordinator's final product, each frame is a sequence of weighted blends of each active choreographer's shadow pose. We compute this product using the *pose dataflow graph*, which dictates the

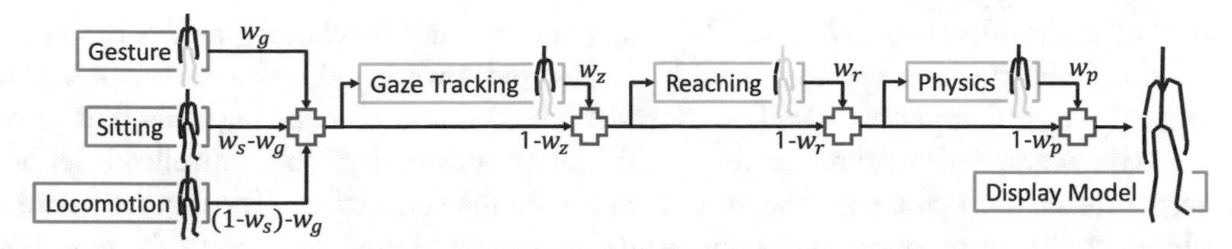

Figure 21.4: A sample dataflow graph we designed for evaluating ADAPT. Generating choreographers appear in blue, transforming choreographers appear in green, and blend nodes appear as red crosses. The final display model node is highlighted in orange. The sitting weight w_s, gesture weight w_g, gaze weight w_z, reach weight w_r, and physical reaction weight w_p are all values between some very small positive ϵ and $1 - \epsilon$.

order of updates and the flow of shadow poses between choreographers. Generators pass data to transformers, which can then pass their data to other transformers, until a final shadow pose is produced, blended with others, and applied to the display model.

Blending is accomplished at certain points in the pose dataflow graph denoted by *blend nodes*, which take two or more input shadows and produce a weighted blend of their transforms. If the weights sum to a value greater than 1, they are automatically normalized.

$$B(\{(\mathbf{P}_i, w_i) : i = 1..n)\}) \longrightarrow \mathbf{P}' \tag{21.1}$$

Designing a dataflow graph is a straightforward process of dictating which nodes pass their output to which other nodes in the pipeline, and the graph can be modified with minimal effort. The dataflow graph for a character is specified by the user during the design and authoring process, connecting choreographers with blend nodes and one another. The weights involved in blending are bound to edges in the graph and then controlled at runtime by commands from the behavior system. The order of the pose dataflow graph roughly dictates the priority of choreographers over one another. Choreographers closer to the final output node in the graph have the authority to overwrite poses produced earlier in the graph, unless bypassed by the blending system. Changing the order of nodes in the dataflow graph will affect these priorities, and so we generally design the graph so that choreographers controlling more parts of the body precede those controlling fewer.

Blended poses are calculated on a per-joint basis using each joint's position vector and orientation quaternion. The weighted average we produce accommodates cases where parts of a shadow's skeleton have been pruned or filtered from the blend (such as an upper-body shadow missing the character's legs). The blend function produces a new shadow pose that can be passed to other transformers, or be applied to the display model's skeleton.

Figure 21.4 illustrates a sample dataflow graph, incorporating generating and transforming choreographers, as well as four blend nodes. Three generating choreographers (blue) begin the pipeline. The gesture choreographer affects only the upper body, with no skeleton information for the lower body. Increasing the value of the gesture weight w_g places this choreographer in control of the torso, head, and arms. The sitting and locomotion choreographers can affect the entire body, and the user controls them by raising and lowering the sitting weight w_S. If w_g is set to $1 - \epsilon$, the upper body will be overridden by the gesture choreographer, but since the gesture choreographer's shadow has no legs, the lower body will still be controlled by either the sitting or locomotion choreographer as determined by the value of w_S. The first red blend node combines the three produced poses and sends the weighted average pose to the gaze tracker. The gaze-tracking choreographer receives an input shadow pose, and applies an offset to the upper body to achieve a desired gaze target and produce a new shadow pose. The second blend node can bypass the gaze tracker if the gaze weight w_z is set to a low value (ϵ). The reach and physical reaction choreographers receive input and can be bypassed in a similar way. The final result is sent and applied to joints of the display model, and rendered on screen. The dataflow graph accommodates the addition of new choreographers in a generalizable fashion, allowing a user to insert new nodes and blend between the poses they create. Rather than designing animation modules to explicitly negotiate, the coordinator seamlessly fades control of parts of the body between arbitrary choreographers in an authorable pipeline.

21.3.3 USING CHOREOGRAPHERS AND THE COORDINATOR

The dataflow graph, once designed, does not need to be changed during runtime or to accommodate additional characters. Instead, the coordinator provides a simple interface comprising messages and exposed blend weights for character animation. Messages are commands (e.g., `SitDown()`) relayed by the coordinator to its choreographers, making the coordinator a single point of contact for character control, as illustrated in Fig. 21.2. In addition to messages, the weights used for blending the choreographers at each blend node in the dataflow graph are exposed, allowing external systems to dictate which choreographer is active and in control of the body (or a segment of the body) at a given point of time. For example, in Fig. 21.4, lowering w_s will transfer control of the body to the locomotion choreographer, while raising its value will give influence to the sitting choreographer. Both choreographers are still manipulating their shadows each update, but only one choreographer's shadow pose is displayed on the body at a given time, with smooth fading transitions between the two where necessary.

For gesturing, we raise w_g, which takes control of the arms and torso away from both the locomotion and sitting choreographers and stops the walking animation's arm swing. Given sole control, the gesture choreographer plays an animation on the upper body, and then is faded back out to allow the walking arm-swing to resume. Since the gesture choreographer's shadow skeleton has no leg bones, it never overrides the sitting or locomotion choreographer, so the lower body will still be sitting or walking while the upper body gesture plays. All weight changes are smoothed over several frames to prevent jitter and transition artifacts. Note that the controllers are never in direct communication to negotiate this exchange of body control. The division of roles between the coordinator and choreographers centralizes character control to a single externally facing character interface, while leaving the details of character animation distributed across modular components that are isolated from one another and can be easily updated or replaced.

Shadow Pose Post-Processing. Since shadow poses are serializations of a character's joints, additional nodes can be added to the pose dataflow graph to manipulate shadows as they are transferred between choreographer nodes or blend nodes. For instance, special filter nodes can be added to constrain the body position of a shadow pose, preventing joints from reaching beyond a comfortable range by clamping angles, or preventing self-collisions by using bounding volumes. Nodes can be designed to broadcast messages based on a shadow's pose, such as notifying the behavior system when a shadow is in an unbalanced position, or has extended its reach to a certain distance. The interface for adding new kinds of nodes to a pose dataflow graph is highly extensible. This affords the user another opportunity to quickly add functionality to a coordinator without directly modifying any choreographers.

21.3.4 EXAMPLE CHOREOGRAPHERS

ADAPT provides a diverse array of character choreographers for animating a fully articulated, expressive virtual character. Some of these choreographers were developed specifically for ADAPT, while others were off-the-shelf solutions used to highlight the ease of integration with the shadow framework. ADAPT is designed to "trick" a well-behaved character control system into operating on a dedicated shadow model rather than the display model of the character, and so the process of modifying an off-the-shelf character control library to a character choreographer often requires modifying only a few lines of code. Since shadows replicate the structure and functionality of a regular character model, no additional considerations are required once the choreographer has been retargeted to the shadow. Note that the choreographers presented here are largely baseline examples. The focus of ADAPT is to allow a user to add additional choreographers, experiment with new techniques, and easily exchange generic choreographers with more specialized alternatives.

Locomotion. ADAPT uses a semi-procedural motion-blending locomotion system for walking and running released as a C# library with the Unity3D engine [105]. The system takes in animation data, analyzes those animations, and procedurally blends them according to the velocity and orientation of the virtual character. We produced satisfactory results on our test model using five motion capture animation clips. Additionally, the user can annotate the character model to indicate the character's legs and feet, which allows the locomotion library to use inverse kinematics for foot placement on uneven surfaces. We extended this library to work with the ADAPT shadow system, with some minor improvements.

Gaze Tracking. We use a simple IK-based system for attention control. The user defines a subset of the upper body joint hierarchy which is controlled by the gaze tracker, and can additionally specify joint rotation constraints and delayed reaction speeds for more realistic results. These parameters can be defined as functions of the character's velocity or pose, to produce more varied results. For instance, a running character may not be permitted to rotate its torso as far as a character standing still.

Upper Body Gesture Animations. We dedicate a shadow with just the upper body skeleton to playing animations such as hand gestures. We can play motion clips on various parts of the body to blend animations with other procedural components.

Sitting and Standing. The sitting choreographer maintains a simple state machine for whether the character is sitting or standing, and plays the appropriate transition animations when it receives a command to change state. This choreographer acts as an alternative to the locomotion choreographer when operating on the lower body, but can be smoothly overridden by choreographers acting on the upper body, such as the gaze tracker.

Reaching. We implemented a simple reaching control system based on Cyclic Coordinate Descent (CCD) [220]. We extended the algorithm to dampen the maximum angular velocity per frame, include rotational constraints on the joints, and apply relaxation forces in the iteration step. During each iteration of CCD (100 per frame), we clamp the rotation angles to lie within the maximum extension range, and gently push the joints back towards a desired "comfortable" angle for the character's physiology. These limitations and relaxation forces are based on an empirical model for reach control based on human muscle strength [256]. This produces more realistic reach poses than naïve CCD, and requires no input data animations. The character can reach for an arbitrary point in space, or will try to do so if the point is out of range.

Physical Reaction. By allocating an upper-body choreographer with a simple ragdoll, we can display physical reactions to external forces. Once an impact is detected, we apply the character's last pose to the shadow skeleton, and then release the ragdoll and allow it to buckle in response to the applied force. By quickly fading in and out of the reeling ragdoll, we can display a physically plausible response and create the illusion of recovery without requiring any springs or actuators on the ragdoll's joints.

SmartBody Integration. To access its locomotion and procedural reaching capabilities, we integrated the ICT SmartBody [240] into ADAPT, using SmartBody's Unity interface and some modifications. Since our model's skeleton hierarchy differed from that of the default SmartBody characters, sample animations had to be retargeted to use on our model. Additionally, our animation interface needed to interact with SmartBody using BML [297]. Since our coordinator is already designed to relay messages from the behavior system, changing those messages to a BML format was a straightforward conversion. Overall, the SmartBody choreographer blends naturally with other choreographers we have in the ADAPT framework, though SmartBody has other features that we do not currently exploit. This process demonstrates the efficacy of integrating other available libraries or external solutions.

21.4 CHARACTER BEHAVIOR

21.4.1 THE ADAPT CHARACTER STACK

At its core, one of the most fundamental problems in interactive character animation is the task of converting simple commands like "reach for that object", or even more sophisticated, cooperative directives like "engage in a conversation," into a series of complicated joint actuations on one or more articulated bodies. ADAPT accomplishes this task with a hierarchy of abstractions comprising what we call the *ADAPT Character Stack*, where one instance of a complete character stack represents one character in ADAPT. The stack is split into four main tiers: Behavior, Actor, Body, and Animation (comprising the Navigation and Coordinator/Choreographer systems described in Section 21.3).

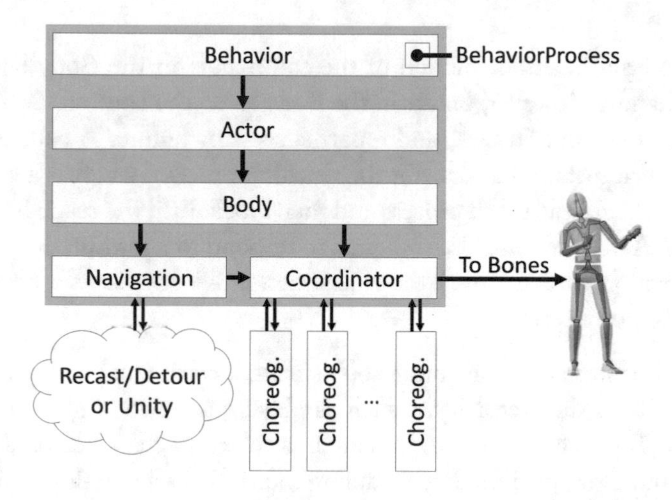

Figure 21.5: The ADAPT character stack.

Commands from each layer of the stack are filtered, converted, and distributed to subcomponents, starting first as behavior invocations, then translating to messages sent to the navigation or animation system, and finally converting into joint angles and blend weights used for posing the character on a frame-by-frame basis. Each layer in the character stack provides a different entry point for technical control over the character. The "Behavioral" layers offer interfaces for controlling the character at a high level, suitable for invocation by behavior trees (Fig. 21.5). The "Mechanical" and "Technical" levels offer more fine-grain control fidelity at the expense of simplicity, and can be invoked to ensure very specific constraints on the character's movements. The bottom layers of the stack are not intended for external access in most circumstances. The exact role of each layer in the character stack is as follows:

Animation (Navigation and Coordinator). This layer provides the lowest-level external access to the character's animation. A component accessing this part of the character stack is concerned with sending messages directly to choreographers (such as to change the reaching target position), or in modifying the blend weights (described in Section 21.3.2) to raise or lower the blend influence of a choreographer.

Body. This layer converts abstract commands like `ReachFor()` into a series of messages passed to the reach choreographer and coordinator to set the reach target and raise the blend weight for the reaching pose. The Body layer is created to encapsulate the pose dataflow graph for a particular character, and assigns more semantic meaning to the blend weights for each blend node in the graph. An ADAPT character's list of capabilities is discussed in more detail in Section 21.4.2.

Actor. This layer is an abstraction of the commands in the Body layer, and provides the same set of commands. However, unlike in the Body layer, the commands in the Actor layer will keep track of the duration of a task, and report success or failure. A call to `ReachFor()` in the Body layer will return instantly and begin the reaching process, whereas a call to `ReachFor()` in the Actor layer will begin the reach process and then block until the reach has succeeded or failed. Commands in the Actor layer are also designed to respond to a termination signal for scheduling, as described in Section 21.5.1. This layer of abstraction is necessary for controlling a character's behavior with behavior trees.

Behavior. This layer contains more sophisticated, contextual commands comprising multiple sequential calls to the Actor layer, such as playing a series of gestures to convey approval in a conversation. The Behavior layer also contains the character's personal behavior tree, and a BehaviorProcess node responsible for scheduling multi-character actions using the ADAPT behavior scheduler described in Section 21.5.1. Unless involved in a multi-actor event, the Behavior layer is responsible for directing the character's goals, and external calls to the Behavior layer are usually concerned with suspending or re-activating a character's autonomy.

21.4.2 BODY CAPABILITIES

In the character stack's Body layer, the navigation and shadow-based character animation system provides a number of capability functions, including:

Commands	Description
ReachFor(target)	Activates the reaching choreographer, and reaches towards a position.
GazeAt(target)	Activates the gaze choreographer, and gazes at a position.
GoTo(target)	Begins navigating the character to a position.
Gesture(name)	Activates the gesture choreographer for the duration of an animation.
SitDown()	Activates the sitting choreographer and sits the character down.
StandUp()	Stands the character up and then disables the sitting choreographer.

Passing an empty target position will end that task, stopping the gaze, reach, or navigation. The locomotion choreographer will automatically react to the character's velocity, and move the legs and arms to compensate if the character should be turning, walking, side-stepping, backpedaling, or running. Note that only sitting and navigating are mutually exclusive. All other commands can be performed simultaneously without visual artifacts.

Adding a New Body Capability

Adding a new behavior capability with a motion component, such as climbing or throwing an object, requires a choreographer capable of producing that motion. Choreographers can be designed to perform animation tasks based on animation data, procedural techniques, or physically driven models. Once the choreographer is developed, the process of adding a new behavior capability to take advantage of the choreographer requires two steps. First, the choreographer must be authored into the pose dataflow graph, either as a generating or transforming node, with appropriate connections to blend nodes and other choreographers. Next, the behavior interface can be extended with new functions that either modify the blend weights relevant to the new choreographer, or pass messages to that choreographer by relaying them through the coordinator. The sophistication of character choreographers varies, but the process of integrating a functioning choreographer into the behavior and animation pipeline for a character is authorable and generalizable.

21.5 CHARACTER INTERACTIONS

Using a character's body capability repertoire, we can produce more sophisticated actions as characters interact with one another and the environment. Authoring complex behaviors requires an expressive and flexible behavior-authoring structure granting the behavior designer reasonable control over the characters in the environment. To accomplish this task, we use parameterized behavior trees (PBTs) [243]. PBTs are an extension of behavior trees [102] that allow them to manage and transmit data within their hierarchical structure without the use of a blackboard.

21.5.1 CHARACTERS INTERACTING WITH EACH OTHER

A useful advantage of PBTs is that they can simultaneously control multiple characters in a single reusable structure called an *event* [262]. Events are pre-authored behavior trees that sit uninitialized in a library until invoked at runtime. When instantiated, an event takes one or more actors as parameters, and is temporarily granted exclusive control over those characters' actions. While in control, an event treats these characters as limbs of the same entity, dispatching commands for agents to navigate towards and interact with one another. Once the event ends, control is yielded to each character's own individual decision processes, which can also be designed using PBTs or with some other technique. Events are a convenient formalism to use for interactions with a high degree of interchange and turn-taking, such as conversations. A conversation event can be authored as a simple sequential or stochastic sequence of commands directing agents to face one another and take turns playing gesture animations or exchanging physical objects.

Figure 21.6 illustrates a sample behavior tree event conducting two characters through a conversation using our action repertoire. The characters, `a1` and `a2`, are passed as parameters to the

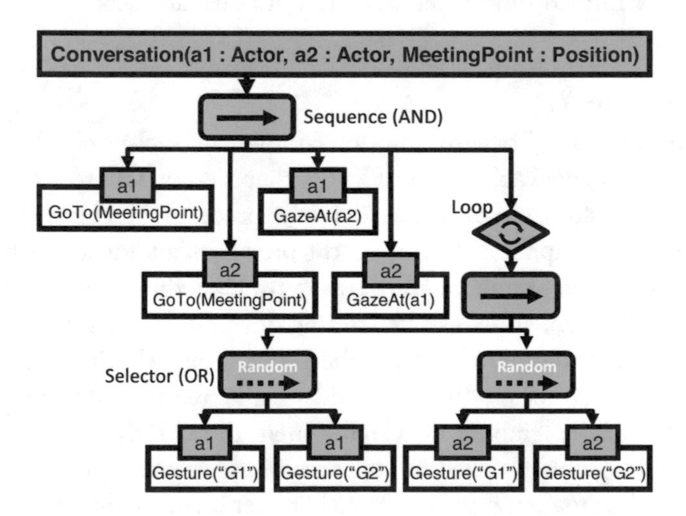

Figure 21.6: A simple conversation PBT event controlling two characters, `a1` and `a1`, with one additional `MeetingPoint` parameter.

tree, along with the meeting position. Using the action interface, the tree directs the two characters to approach one another at the specified point, face each other, and alternatively play randomly selected gesture animations. The gesturing phase lasts for an arbitrary duration determined by the configuration of the loop node in the tree. After the loop node terminates, the event ends, reporting success, and the two characters return to their autonomous behaviors. Note that this tree can be reused at any time for any two characters and any two locations in the environment in which to stand. This framework can be exploited to create highly sophisticated interactions involving groups or crowds of agents (e.g., [266]), and its graphical, hierarchical nature makes subtrees easier to describe and encapsulate.

The ADAPT Behavior Scheduler

ADAPT provides a fully featured scheduler for managing and updating both the personal behavior trees belonging to each character and higher-level event behavior trees encompassing multiple characters. Four basic principles in behavior design enable the scheduler to work effectively for orchestrating the behavior in an ADAPT environment.

PBT Clock. PBTs in ADAPT operate on periodic clock ticks, where each tree refreshes itself, evaluates its current state, sends messages through the character stack, and transitions to its next node if necessary. The ADAPT scheduler keeps track of all of the active trees in the environment, both personal trees for individual characters, and event trees for multi-character interactions, and ticks them 30 times per second.

Character Suspension. Each character in ADAPT owns a "BehaviorProcess" object in its behavior layer, which maintains that character's state with respect to autonomy. A character will not receive ticks to its personal tree if it has been suspended and placed under the control of a multi-character event tree.

Behavior Termination. PBTs in ADAPT are designed to respond to a termination signal. Termination signals can come at any time, and instruct a tree to interrupt its current action and end. The result of a termination signal can last multiple PBT clock ticks, so that poses such as reaching can be smoothly faded out before the tree reports that it has completed termination.

Event Priority. Multi-character event trees are all assigned a priority value, where all character personal trees have minimal priority. When a character receives an event tree with a priority higher than its current running tree, the scheduler sends a termination signal to that character's tree. Once the termination completes, the scheduler places that character into suspension. When all of the characters involved in the event are terminated and suspended, the event can begin ticking and dispatching commands to them. Once the event terminates, the characters return from being suspended.

These four concepts allow very direct control over groups of characters in the environment, with smooth transitions between drastically different tasks. Since trees can be cleanly terminated at any point in their execution, groups of characters involved in wandering or conversing with one another in an environment could very quickly activate a new, higher priority event tree to respond quickly to an event such as a loud noise or a fire. Each tree operates as its own cooperatively multi-threaded process with no direct inter-communication. Since a character is only ever controlled by one tree at a time, this presents an opportunity for parallelization in future work.

21.5.2 CHARACTERS INTERACTING WITH THE ENVIRONMENT

Using the same for principles for the behavior scheduler allows us to more easily implement smart objects into ADAPT. A smart object's affordances can be encoded sequentially in a manner similar to that of a behavior tree. Smart objects receive ticks from the scheduler clock, and will block until reporting either success or failure. This provides a convenient interface for interaction with a behavior tree. For example, when a character wants to sit in a chair, the behavior tree invokes the chair's "sit" affordance with the character as a parameter. From that point, the tree will divert any ticks it receives from the scheduler clock to the smart object, which temporarily takes control of the character and directs it to approach and sit properly on the chair. Once the smart object chair determines that the character has succeeded in sitting down, it will report success to the behavior tree responsible for the character's behavior, so that the tree can move on to other actions. Parallel nodes in the tree can be used to synchronize actions, allowing a character to gesture or gaze at a target while still approaching and sitting.

Smart objects represent the primary means of interaction with the environment, and are useful for a wide array of other interaction tasks. For example, in our conversation behavior tree in Fig. 21.6, we instruct both characters to approach a meeting point before talking to one another. This can be easily accomplished with a smart object waypoint that temporarily takes control of the two characters, directs them to approach a point using the navigation system, and reports success back to the behavior tree once the characters are within close proximity and oriented towards one another.

21.6 RESULTS

Using ADAPT, we can create a character that can simultaneously reach, gaze, walk, and play gesture animations, as well as activate other functionality like sitting and physically reacting to external forces. ADAPT characters can intelligently maneuver an environment, avoiding both static obstacles and one another. These features are used for authoring sequences like exchanging an object between actors, wandering while talking on a phone, and multiple characters holding a conversation.

21.6.1 MULTI-ACTOR SIMULATIONS

We author a simple narrative to demonstrate the features of ADAPT. Events, once active, can be successfully executed or interrupted by other triggers due to dynamic events, or user input. This produces a rich interactive simulation where virtual characters can be directed with a high degree of fidelity, without sacrificing autonomy or burdening the user with authoring complexity.

In the beginning, an event ensues where a character is given a phone and converses while wandering through the scene, gazing at objects of interest. The phone conversation event successfully completes and the character hands back the phone. Spotting nearby friends invokes a conversation, which is an extension of the event illustrated in Fig. 21.6. The conversation is interrupted when a child throws a ball at one of the characters. The culprit flees from the scene, triggering a chasing event where the group runs after the child. The chase fails as the child is able to escape through a crossing crowd of characters, which are participating in a group event to navigate to the theater and find a free chair to sit.

21.6.2 COMPUTATIONAL PERFORMANCE

ADAPT supports approximately 150 agents with full fidelity at interactive frame rates on a standard desktop computer for a single-threaded application. Figure 21.7 displays the update frequency for the animation and navigation system (for our scenes, the computational cost of behavior was negligible). This varies with the complexity of the choreographers active on each character.

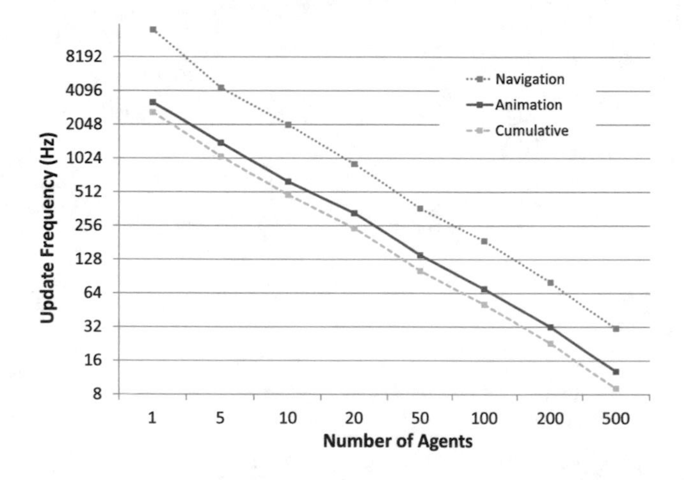

Figure 21.7: Update frequency for the character animation and navigation components in ADAPT.

The ADAPT animation interface and the pose dataflow graph has little impact on performance, and the blend operation is linear in number of choreographers. Each joint in a shadow is serialized with 7 4-byte float values, making each shadow 28 bytes per joint. For 26 bones, the shadow of a full-body character choreographer has a memory footprint of 728 bytes. For 200 characters, the maximum memory overhead due to shadows is less than 1 MB. In practice, however, most choreographers use skeleton subsets with just the upper body or only a limb, making the actual footprint much lower for an average character.

Separating character animation into discrete modules and blending their produced poses as a post-processing effect also affords the system unique advantages with respect to dynamic level-of-detail (LOD) control. Since no choreographer is architecturally dependent on any other, controllers can be activated and deactivated arbitrarily. Deactivated controllers can be smoothly faded out of control at any time, and their nodes in the dataflow graph can be bypassed using the existing blend weights. This drastically reduces the number of computed poses, and conserves processing resources needed for background characters that do not require a full repertoire of actions. The system retains the ability to re-activate those choreographers at any time if a specific complex action is suddenly required. Since choreographers are not tightly coupled, no choreographer needs to be made aware of the fact that any other choreographer has been disabled for LOD purposes.

CHAPTER 22

Event-centric Planning for Narrative Synthesis

Alexander Shoulson, Max Gilbert, Mubbasir Kapadia, and Norman I. Badler

Complex virtual worlds with sophisticated, autonomous virtual populations are an essential tool in developing simulations for security, disaster prevention, interactive training experiences and entertainment. Directed interactions between virtual characters help tell rich narratives within the virtual world and personify the characters in the mind of the user. However, creating and coordinating fully fledged virtual actors to behave realistically and act according to believable goals is a significant undertaking. This difficulty is amplified when we design these characters to cooperate or compete according to their personal motivations and needs, as two autonomous characters interact in a far more intricate fashion than would a character and a prop in the environment. In order to steer the trajectory of the narrative in a meaningful fashion, we need a virtual director capable of making narrative decisions based on an understanding of the world and its characters, including those characters' abilities and individual goals. This volume of information is too large to be managed in real-time, requiring a level of abstraction to temper the complexity of a rich world and the emergent nature of its inhabitants.

Traditional stories are usually written in a form that emphasizes parsimony. Even the most prosaic description of characters engaging in a conversation is not likely to concern itself with details like the characters approaching and orienting towards one another, or listeners changing their posture to affect attention towards the speaker, unless these details deviate from the reader's

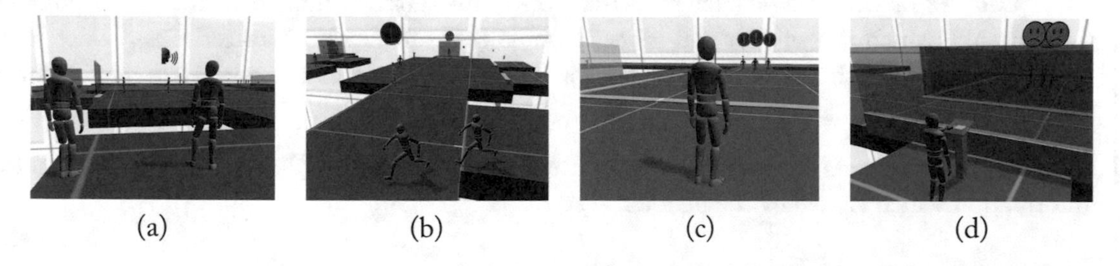

(a) (b) (c) (d)

Figure 22.1: Characters exhibiting two cooperative behaviors. Two actors hide while a third lures a guard away (a), then sneak past once the guard is distracted (b). An actor lures the guards towards him (c), while a second presses a button to activate a trap (d).

expectation in a meaningful way or contribute artistic value. Similarly, a planning controller in a virtual world designed for narrative impact should not be concerned with details more significant to the execution of an animation than with the story being presented. Still, these mechanical elements cannot simply be excluded—as minor as the action may be, two actors in a conversation must still visibly approach and orient towards one another. To collapse a very large, mechanically oriented character action space into a series of tools available to a narrative planner, we define and use a domain of *events*: dynamic and reusable behaviors for groups of actors and objects that carry well-defined narrative ramifications. The ADAPT system provides a suitable foundation for expressing and executing such events.

Using events as an encapsulation of the traditional atomic action planning domain affords an author the following advantages:

Authoring Fidelity. While events can be nondeterministic, they are generally authored with a very specific trajectory of actions. An event will execute a series of actions in a predetermined fashion, so long as the event itself is successful. This allows an author to create emergence in the overarching narrative structure of a simulation between events while still ensuring that when an event occurs, it will proceed as intended. Since the results of a planner are difficult to verify in practice for large domains, being able to verify the behavior of each event prior to release tames any tendencies for the system to produce undesired effects.

Synchronization. Events can be used to coordinate very sophisticated multi-actor behaviors like conversation or group activities that require precisely timed animations or procedural character controllers. Rather than depending on a planner to organize the subtasks of a compound character interaction, a pre-authored event can be written to properly time each action's execution between any number of actors involved.

Efficiency. An event takes full control of its participants and treats them as if they were limbs of the same entity. This allows multiple events to control numerous diverse sets of characters in parallel, and decouples the complexity of the planning domain from the number of agents in the world. Rather than searching through all possible actions a character can perform on all other objects and characters in the world, the search is limited to the set of possible events, which is a smaller and more tractable domain. Event characteristics replace local intelligence in the character, allowing for "lightweight" agents and hence a greater number of potential scenario participants.

We author these events using Parameterized Behavior Trees (PBTs) [243] with additional meta-information which is used by the planner.

22.1 PROBLEM DOMAIN AND FORMULATION

For planning in the space of narrative, we use a generalized best-first planner operating on a problem domain defined as $\Sigma = \langle \mathbf{S}, \mathbf{A} \rangle$ where $\mathbf{S}$ is the combined state of all objects and actors in the environment, and **A** is the space of authored events. A particular problem instance is defined as $\mathbf{P} = \langle \Sigma, S_{\text{start}}, S_{\text{goal}} \rangle$ where $S_{\text{start}}, S_{\text{goal}} \in \mathbf{S}$ are the initial and desired world configurations. The planner generates a sequence of event transitions $\Pi(S_{\text{start}}, S_{\text{goal}})$ that leads the simulation from S_{start} to S_{goal}.

22.1.1 STATE SPACE

Each object in the world **W** is described as $o = \langle c, s \rangle \in \mathbf{W}$ where c is a controller and s is the object's state. We define s as $s = \langle r, p, d \rangle$ where r is a role index number, p is a 3-vector for world position, and d is a high-level description of an object as an unsigned bitmask. For example, a door will have a state description where "opened" and "locked" correspond to 0 or 1. The role index number defines an object's archetype, so that a character is different from a chair, and thus fills a different role in the narrative.

The overall state of the world is defined as the compound state of all of the objects in that world, $S = \{s_1, s_2, \ldots, s_n\}$ where s_i is the state of o_i. Objects can be actors, which have their own autonomy, or props, which cannot act unless involved in an event that directs their behavior. States are defined and hashed by name. For overarching world information describing the global state of the simulation, a special world control object with no physical representation can be read and written to as if it were a prop in the world. This is useful for maintaining narrative trajectories and story arcs.

The composite state presented to the planner is an encapsulation of each object's complete low-level state. Props and actors contain a great deal of information concerning their animation and procedural pose controllers, such as where in the world they should be gazing or reaching, or how near their current gesture animation is to completion, but this information is managed at the event level rather than at the planning level. A behavior author is responsible for establishing the proper connections, so that when a door plays its closing animation the corresponding state in its high-level description is also set to `Closed`. Additionally, entire objects may not be accounted for in the composite state space of the planner. Independent actors out of the planner's control may serve as confounding agents by interacting with objects tracked by the planner and altering their state in unforeseen ways. The most obvious example of such an agent would be a human user, but virtual actors acting with their own autonomy may also fill this role. Because these confounding agents may alter states critical to the progress of a plan, the planner is capable of monitoring required conditions and replanning when necessary.

Some example high-level states for objects in the world are as follows:

- Door: [`Open`, `Locked`, `Guarded`]
- Character: [`HasKey`]
- Guard (extends from Character): [`Trapped`, `Dazed`]

22.1.2 ACTION SPACE

Unlike traditional STRIPS or GOAP planning [64, 201], our planner does not operate in the space of each actor's individual capabilities. Rather, our system's set of actions is taken from a dictionary of *events*, authored parameterizable interactions between groups of characters and props. Because of this disconnection between character behaviors and the action space for the planner, actors can have a rich repertoire of possible actions without making an impact on the planner's computational demands. Two characters involved in an interaction could have dozens of possible actions for displaying diverse, nuanced sets of gesticulations. Where a traditional planner would need to expand all of these possibilities at each step in the search, our planner only needs to invoke the conversation event between the two characters and let the authored intelligence of the event itself dictate which gestures the characters should display and when.

Formulating Narrative Events

Events are pre-authored dynamic behaviors that take as parameters a number of actors or props as participants. When launched, an event suspends the autonomy of any involved object and guides those objects through a series of arbitrarily complex actions and interactions. Events are well-suited for interpersonal activities such as conversations, or larger-scale behaviors such as a crowd gathering in protest. Each event is temporary, possessing a clear beginning and at least one end condition. When an event terminates, autonomy is restored to its participants until they are needed for any other event. Events are represented by Parameterized Behavior Trees (PBTs), as described in the previous section. Each event is defined as

$$e = \langle t, c, \mathbf{R} = (r_1, \dots, r_m), \phi : \mathbf{R} \times \mathbf{W}^m \to \{0, 1\}, \delta : S \to S' \rangle$$

where the PBT t contains event behavior, c is the event's cost, the role list $\mathbf{R}$ defines the number of objects that can participate in the event and each participant's required role index number, the precondition function ϕ transforms the list of m roles and a selection of m objects from the world into a true or false value, and the postcondition function δ transforms the world state as a result of the event. The precondition function will return true if and only if the given selection of objects satisfies both the roles and authored preconditions of the event.

Roles for an event are defined based on an object's narrative value. A conversation may take two human actors, whereas an event for an orator giving a speech with an audience might take a speaker, a pedestal object, and numerous audience members. Preconditions and postconditions define the rules of the world, e.g., a character can pull a lever if the lever can be reached from the character's position. Figure 22.2 illustrates the pre- and post-conditions for an *UnlockDoor* event. This event metadata would be accompanied with a PBT that instructs the character to approach the door and play a series of animations illustrating the door being unlocked.

An event's cost is partially authored and partially generated. An author designing an event where a prisoner obtains a key from a guard may specify that sneaking up to the guard and stealing the key may be less costly than assaulting the guard outright to obtain the key. Other costs, such

```
Event UnlockDoor(Prisoner : a, Door: d) {
  Precondition φ:
   Closed(d) ∧ ¬Guarded(d)
   ∧ Locked(d) ∧ CanReach(a,d);
  Postcondition δ:
   ¬Locked (d)
}
```

Figure 22.2: Event definition for *UnlockDoor*.

as the distance to a navigation goal where an event takes place, are more mechanical and can be automatically determined by the system at runtime. Since we are planning in a narrative space, it is important to author costs so that the planner picks events that are not necessarily the most efficient, but rather are the most consistent with the story. A more appropriate way to think of an event's cost would be to consider the risk to a viewer's suspension of disbelief, rather than to consider the theoretical energy expended by the participants to accomplish the goal. Ultimately there are no fixed rules to determine cost: all costs are relative, and the responsibility of the PBT author.

22.1.3 GOAL SPECIFICATION

Goals can be specified as: (1) the desired state of a specific object, (2) the desired state of *some* object, or (3) a certain event that must be executed during the simulation. These individual primitives can be combined to form narrative structures suited for the unfolding dynamic story. When searching through **A** for an event sequence that satisfies the narrative, the planner uses preconditions to determine possible events, and postconditions to generate future world states after the execution of one or more events. Requirements on the desired state of *some* object are translated into the composite world space by adding multiple satisfiable conditions to detect an end state. The goal stipulating that an event must occur places a hard constraint on any generated plan, rejecting branches even if they have achieved every other goal with minimal cost.

Goals are a manner of specifying the narrative structure of the virtual world. Much of the emergent story using this system will occur when the planner attempts to achieve a narrative objective despite the efforts of confounding agents in the world (including but not limited to the player). It is important to view goals not as means to solve a task, but as the desired climax or conclusion of a story arc. Sequences of goals could be achieved in order to enforce structure on a story that can deviate from expectation due to user interaction. The selection of goals is currently author-specified, though intelligent dynamic selection of goals would allow more user freedom and is a topic we intend to explore in the future.

22.2 PLANNING IN EVENT SPACE

Our system uses a generic best-first planner to produce sequences of events that satisfy a given narrative. The set of permissible events at each search node $I \in \mathbf{I}$ determines the transitions in the search graph. A valid instance of an event e is defined as $I_e = \langle e, O \in \mathbf{W}^{|\mathbf{R}_e|} \rangle$ with $\phi_e(\mathbf{R}_e, O) = 1$. This implies that the event's preconditions have been satisfied for a valid set of participants O, mapped to their designated roles $\mathbf{R}_e$.

The transitions $\mathbf{I}$ are used to produce a set of new states: $\{S' | S_e = \delta(S, I_e) \forall I_e \in \mathbf{I}\}$ by applying the effects of each candidate event to the current state in the search. Cost is calculated based on the event's authored and generated cost, and stored alongside the transition. The planner generates a sequence of event transitions $\Pi(S_{\text{start}}, S_{\text{goal}})$ from the S_{start} to S_{goal} that minimizes the aggregate cost:

$$\Pi(S_{\text{start}}, S_{\text{goal}}) = \underset{\{I_i | 0 \leq i < n\}}{\text{argmin}} \sum_{i=0}^{i<n} c(S_i, I_i)$$

In order to curb the combinatorial cost of this search process, we divide actors and objects into role groups. Though there may be many objects in the world, very few of them will fill any particular role, which greatly reduces the size of the event instance list. Checking an object's role is a very fast operation which we use for filtering before the more costly operation to validate the event's preconditions on its candidates. The maximum branching factor of this technique will be $|\mathbf{E}|(max_{e \in \mathbf{E}} |\mathbf{R}_e|)^{max_{r \in \mathbf{R}} |\ell_r|}$, and the effective branching factor can be calculated by taking average instead of maximal values. Note that this grows independently of an actor or object's individual capabilities, allowing characters to exhibit rich and varied behaviors without affecting the planner's overhead.

22.3 RUNTIME AND SIMULATION

While the planner is responsible for generating sequences of events to produce a narrative, a second component is needed to instantiate those events in a proper order and manage their execution. To do so, we use an event scheduler that monitors a queue of pending events and a list of currently running event behavior trees. The scheduler is responsible for selecting which events to instantiate in order, and for monitoring the world state to detect when preconditions of an event in the plan have been invalidated, necessitating a re-plan.

22.3.1 EVENT LOADING AND DISPATCH

Unlike a contained sequence of actions in a single problem domain, our system presents a fully fledged simulation with many independent actors and moving components. While the story plan itself is a strict sequence, we use a specialized event scheduler that enables us to dispatch multiple events simultaneously if they are not in competition for resources. This makes the simulation appear more lifelike by maximizing actor activity and preventing long periods of time where characters idly wait for the next step in the plan. Multiple discrete "threads" of a story can progress simultaneously in different areas of the world to present a compelling narrative that is driven by the démarches of its characters.

The product of a generated plan is a sequence of events and lists of objects from the world that fill the parameters of those events. Despite the planner's ordering, events using disjoint sets of resources can be safely executed in parallel. Upon receiving a plan, the scheduler queues the plan's event sequence and prepares to dispatch those events to the planner's selected participants. At each scheduler update, the scheduler updates all of the currently running events, and then iterates through the list of queued events to launch those events for which their participants are all available and their preconditions are satisfied. Scheduler updates occur at a fixed frequency and also handle tasks such as cleaning up completed events and restoring autonomy to their participating actors.

In order to maximize the amount of synchronization that can occur, authors of events can wrap complex sequences of actions into *compound events*. Compound events appear to the planner as one large event that takes a group of participating objects. However, when placed on the scheduler's event queue, the compound structure is broken down into a number of component sub-events that each take fewer parameters than the whole. As an example, in a simplified version of our scenario in which three characters escape from a prison, we illustrate a room with a simple puzzle in Fig. 22.3. The three main characters enter the room from the left and must exit on the right. Buttons B1 and B2 open the door at the bottom of the diagram, and must be pushed simultaneously, while button B3 opens the door at the top of the diagram. The pathway to the room with buttons B1 and B2 are blocked by a guard G. The event(s) to solve this could be authored in a number of ways, including one large event that takes all seven parameters and walks the characters through the problem.

Regardless of how the solution appears to the planner, however, the events can be broken down for parallel execution. Figure 22.4 illustrates the resources used for the various sub-events to solve the puzzle. While the events in the sequence may be predicated on state transitions, they may still partially overlap. As character A distracts the guard, characters B and C can press the buttons to open the bottom door. Then A can escape from the guard, press B3, and all three characters can escape simultaneously. Compound events are another tool available to authors to control the branching factor of the planning process, allowing the planner to only see a very coarse-grained event such as "escape from room", without sacrificing qualities like synchronized cooperative behaviors.

22.3.2 HANDLING DYNAMIC WORLD CHANGES

The scheduler receives a notification when any external object in the world, including a user's avatar, has unexpectedly changed the state of the environment. If the change violates any of the currently running events, then those events are immediately terminated, the event queue is

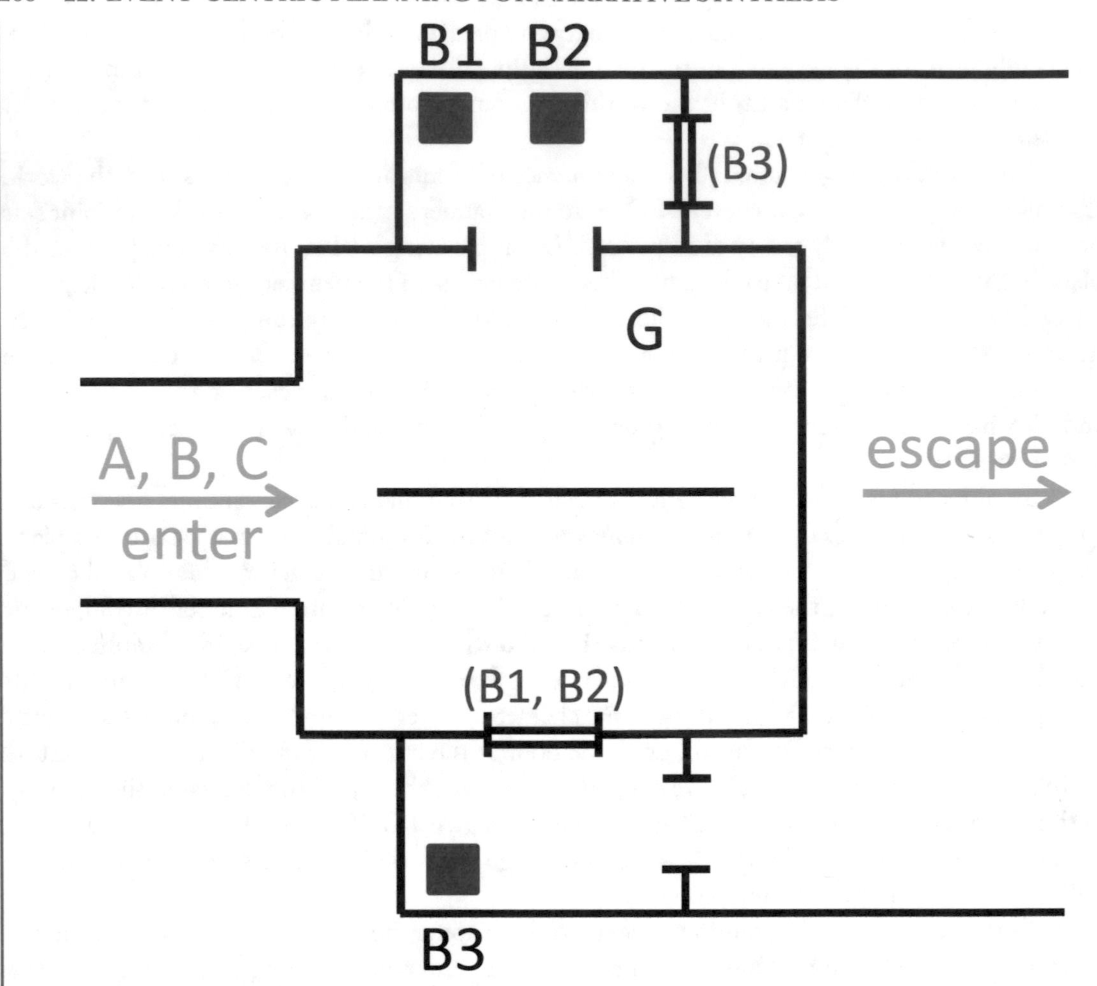

Figure 22.3: Map of the environment for the distract-and-escape scenario.

cleared, and the planner is instructed to re-plan from the current affected world state. If the change does not violate any of the current preconditions, then the scheduler forward-simulates the events in the queue and ensures that given each sequential world state, the next event in the queue can be reached. If any event in the queue is detected as unreachable due to the unexpected change, then the queue is emptied and the planner generates a new sequence of events. This allows the planner to respond to changes from either a user or adversarial virtual characters acting in opposition to the goals of the story. Further work on plan and story repair [167, 206] may be suitable for extending this planning environment.

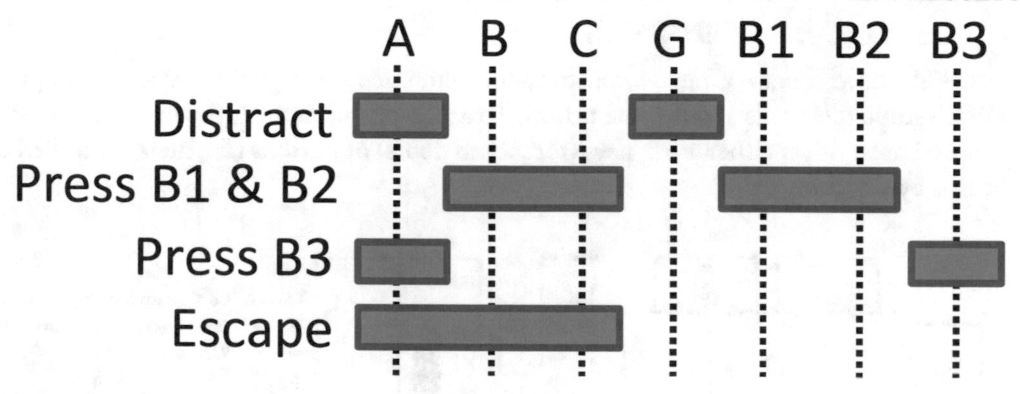

Figure 22.4: Synchronization breakdown of the distract-and-escape scenario.

22.3.3 INTELLIGENT AMBIENT CHARACTER BEHAVIOR

In virtual environments with many actors, we do not expect every actor to be continuously involved in events as the story progresses. At times, characters will be waiting for others to perform tasks and advance the narrative sequence. This can produce periods of downtime where actors are not explicitly given a task to perform by the story planner. Fortunately, the event-centric behavior paradigm allows for characters to maintain their own autonomy when not involved in an event. We extend this functionality so that characters not only maintain a level of activity when waiting for their next story action, but act with an awareness of their current role in the narrative.

Objects in our world are annotated with special *observation points*, which serve as placeholder geometry where a character can safely stand and watch that particular object while it is in use. When a character is idle, we determine which event from the scheduler that character will be involved in next. Given the event, we pick one of the non-actor objects scheduled to be involved in that event, and direct the character to stand in one of that object's observation points as long as the character can safely do so. This produces realistic anticipatory behavior such as a character standing by a door while waiting for another character to unlock and open it. The characters can then be co-opted at any time once their event starts to continue advancing the narrative.

22.4 RESULTS

We author a narrative-driven "prison break" scenario, where prisoner characters need to overcome a number of hurdles including locked doors, traps, alarms, and adversarial guards in order to exit their cells. Section 22.4.1 discusses the virtual environment we designed for the scenario, Section 22.4.2 describes the objects within the virtual world and their state descriptions, Section 22.4.3 describes the process of authoring multi-actor events, Section 22.4.4 describes the generated narratives, and Section 22.4.5 discusses our system's ability to react in real-time to both helpful and confounding user intervention.

22.4.1 ENVIRONMENT DESIGN

Figure 22.5 shows a map of the demonstration environment. The goal is for all of the prisoners to safely escape their cells and flee the prison. To accomplish this goal, the prisoners must evade guards and open doors either using keys (for locked doors) or buttons (for their controlled doors, indicated by matching color).

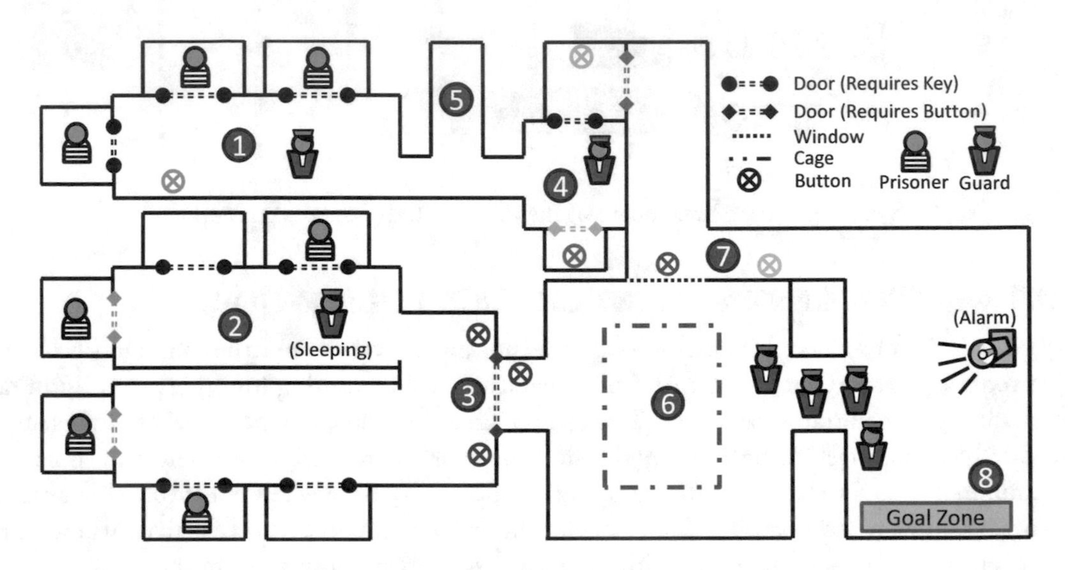

Figure 22.5: Problem definition map for the "prison break" scenario. Prisoners must escape from their cells and escape from the complex by unlocking doors and evading guards.

The map is annotated with the following points of interest (red circles): (1) the north cellblock contains three prisoners in cells, one guard with a key, and a button that opens two doors in the south cellblock (orange doors), (2) the south cellblock contains a sleeping guard with a key and four locked prisoners, (3) the exit to the cellblock is controlled by two buttons on the interior that must be pressed simultaneously to open (dark blue) or one button on the exterior, (4) the exit from the north cellblock is blocked by another guard, and requires three doors to be opened in sequence using a key and then two buttons (light blue, purple), (5) a hiding spot for prisoners to evade the guards in the north cellblock, (6) a cage that rises from the floor and can trap prisoners or guards caught inside, (7) a control room with buttons controlling the cage (red) or the alarm at the exit (green) that will attract guards to it when activated, and (8) the goal zone, representing the exit from the prison.

22.4.2 OBJECT STATE DESCRIPTION

Our world state comprises instances of eight object types, each with their own individual state.

Object	State Fields
Door	Closed, Locked, Guarded, Room1, Room2
Prisoner	HasKey, CurrentRoom, Position
Room	Guarded, AdjacentRoom1, AdjacentRoom2
CageTrap	Active
Button	Active, ControlledObject
Guard	HasKey, Trapped, Dazed, Position
Alarm	Active
Key	(no state information)

The state of an object determines the conditions for its use. For example, a door with a guard standing near it is considered guarded, and cannot be safely opened or unlocked without first removing or disabling the guard. The same applies for guarded rooms, which cannot be safely entered. Buttons are linked to the objects they activate, controlling doors, the cage trap, or the alarm. Guards are unable to move or react when dazed or trapped, and locked doors can only be opened by characters that have a key. For special cases like a door that can only be opened by two simultaneous button presses, we use proxy objects that comprise the state of two or more objects in the world.

22.4.3 AUTHORED EVENTS

Table 22.1 illustrates the major events used in our planning scenario, with their parameters, preconditions, postconditions, and a summary of their behavior trees. Pre- and post-conditions involving character positions are omitted for clarity, but are handled, in general, with a system of room adjacency and world annotations in the form of waypoints. These events were authored as generic reusable structures and placed in a library available to the planner. The process of introducing new behavior is straightforward, and can simply be accomplished by adding a new event that manipulates the objects in the world. By using events, we were able to add a rich variety of behavior to the world with a minimal degree of authorial burden. These events, such as the six-actor event to trap four guards in a cage, allow for rich multi-character interactions without causing a combinatorial growth in the planning branching factor.

22.4.4 GENERATED NARRATIVE

In our experiments, we generated four different narrative plans. The goal of each simulation is for each prisoner to be able to reach the exit location of the prison map. The initial configuration of the world matches that of Fig. 22.5, with prisoners in their cells, and guards with keys guarding the prison exits.

Figure 22.6 provides a storyboard of the first generated narrative plan from the default environment. The plan begins with a prisoner in the north cellblock (area (1) in Fig. 22.5) luring the guard into his cell, assaulting him, and stealing his key (a). Afterwards, that prisoner frees

Table 22.1: Summary of major events used in our narrative scenarios. Note that events may invoke other sub-events in the library during their execution

```
Event DistractGuard(Guard : g; Door : d;
    Prisoner : a, b, c; Waypoint : u, v) {
  Precondition ϕ:
    Guarded(d) ∧ CanReach(a,g) ∧ CanReach(b,u) ∧ CanReach(c,v);
  Postcondition δ:
    ¬Guarded (d);
  Behavior Summmary:
    b hides at u
    c hides at v
    a draws away g
}
```

```
Event EscapeCell(Guard : g; Door : d;
    Prisoner : a) {
  Precondition ϕ:
    CanReach(g,d) ∧ HasKey(g)
    ∧ Closed(d) ∧ Locked(d);
  Postcondition δ:
    HasKey(a) ∧ ¬HasKey(g)
    ∧ Trapped(g) ∧ Locked(d);
  Behavior Summmary:
    a calls g
    OpenDoor(g,d)
    a dazes g
    StealKey(g,a)
    a closes and locks d
}
```

```
Event SoundAlarm(Guard : g, h, i, j;
    Alarm : a; Button : b; Prisoner : p) {
  Precondition ϕ:
    CanReach(p,b) ∧ Controls(b,a)
    ∧ ¬Active(a);
  Postcondition δ:
    Active(a);
  Behavior Summmary:
    p presses b
    g, h, i, j approach a
}
```

```
Event StealKey(Guard : g; Prisoner : a) {
  Precondition ϕ:
    CanReach(a,g) ∧ HasKey(g)
    ∧ IsDazed(g) ∧ ¬HasKey(a);
  Postcondition δ:
    HasKey(a) ∧ ¬HasKey(g);
  Behavior Summmary:
    a approaches g
    a reaches g to take key
}
```

```
Event OpenDoor(Agent : a; Door : d) {
  Precondition ϕ:
    CanReach(a,d) ∧ Closed(d)
    ∧ ¬Locked(d);
  Postcondition δ:
    ¬Closed(d);
  Behavior Summmary:
    a approaches d
    a opens d
}
```

```
Event ExchangeKey(Prisoner : a, b) {
  Precondition ϕ:
    CanReach(a,b) ∧ HasKey(a)
    ∧ ¬HasKey(b);
  Postcondition δ:
    HasKey(b) ∧ ¬HasKey(a);
  Behavior Summmary:
    a approaches b
    a gives b the key
}
```

```
Event TrapGuards(Guard : g, h, i, j;
    Prisoner : a, b; Button : u; CageTrap : t;
    Waypoint : w) {
  Precondition ϕ:
    CanReach(a,w) ∧ CanReach(b,u)
    ∧ OnPathTo((g..j),w,t) ∧ ¬Active(t)
    ∧ Controls(u,t);
  Postcondition δ:
    Trapped(g) ∧ Trapped(h)
    ∧ Trapped(i) ∧ Trapped(j)
    ∧ Active(t);
  Behavior Summmary:
    a approaches w
    a calls out to g, h, i, j
    g, h, i, j approach a
    b presses u
    t traps g, h, i, j

}
```

```
Event TrapGuardsAlarm(Guard : g, h, i, j;
    Prisoner : a, b; Button : u, v;
    CageTrap : t; Alarm : x) {
  Precondition ϕ:
    CanReach(a,u) ∧ OnPathTo((g..j),x,t)
    ∧ ¬Active(t) ∧ Controls(u,t)
    ∧ CanReach(b,v) ∧ Controls(v,x)
    ∧ ¬Active(x);
  Postcondition δ:
    Trapped(g) ∧ Trapped(h)
    ∧ Trapped(i) ∧ Trapped(j)
    ∧ Active(x) ∧ Active(t)
  Behavior Summmary:
    SoundAlarm(g,h,i,j,x,v,a)
    TrapGuards(g,h,i,j,a,b,t)
}
```

the others in the north cellblock, and presses the button (b) to release the south cellblock (orange button, Fig. 22.5). Two prisoners are released (c) in the south cellblock, and one steals a key from the sleeping guard (d). Meanwhile, in the north cellblock, a prisoner lures a guard away from the door (e, f) so that two other prisoners can escape by pressing the correct buttons (g) (area (4) in Fig. 22.5). They then shut the door and lock the guard behind them (h). In the south cellblock, two prisoners simultaneously press the door buttons (i) to open the door (dark blue, area (3) in Fig. 22.5). A prisoner runs out into the cage room (area (6) in Fig. 22.5), and attracts the guards to him (j). Another prisoner presses the button to trap the guards (k). Afterwards, all seven prisoners escape (l).

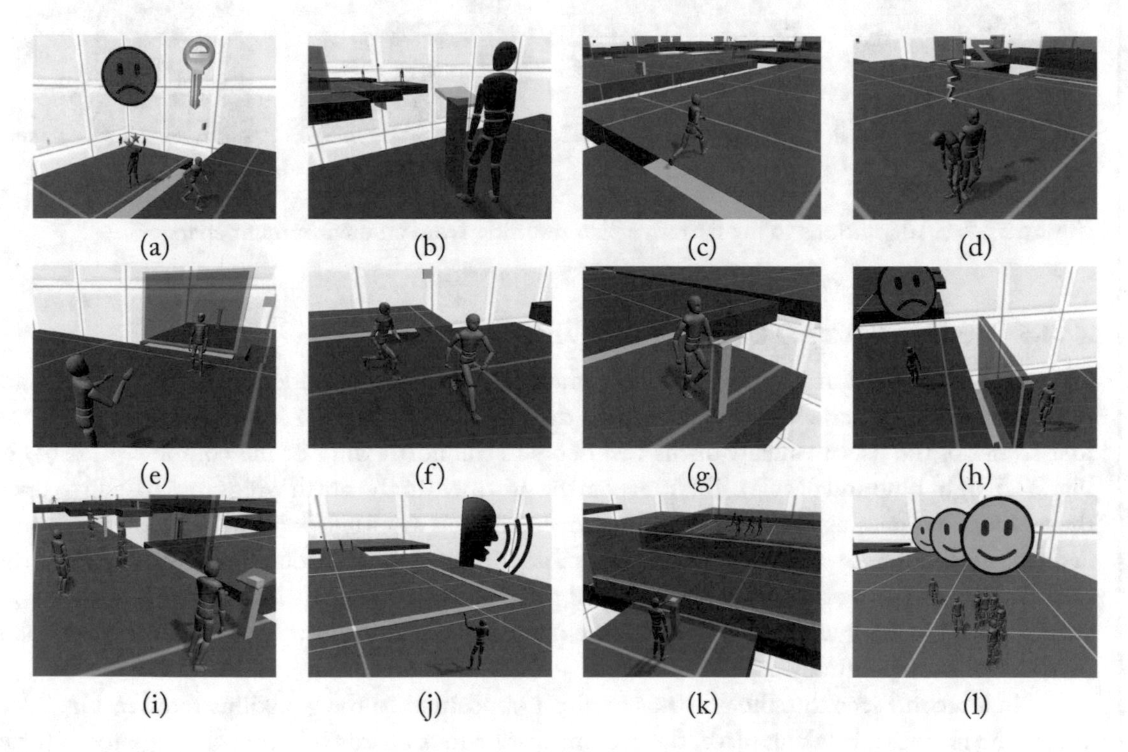

Figure 22.6: Storyboard for the first generated narrative plan with no intervention.

We altered the environment by disabling the button in the north cellblock (near (1) in Fig. 22.5, orange), which prevented the two doors in the south cellblock (orange) from being opened by the north wing prisoners. This one change to the world state produced a drastically different narrative experience.

The second narrative begins in the same way as the first, with the north cellblock prisoner stealing a key from a guard and releasing the other two prisoners (Fig. 22.6(a)). However, with the button disabled, the south cellblock prisoners could not be released, so the events featured in Fig. 22.6(c, d, i) do not occur. Instead, an alternative sequence of actions occurs, displayed in

Fig. 22.7. To lure the guards away from the gate room exit, a prisoner from the north cellblock activates the alarm (m). This attracts the guards to it so that a prisoner can sneak past them (n) and enter the south cellblock. After stealing a key from the sleeping guard (as in 22.6(d)), that prisoner releases the other south cellblock prisoners (o), and lures the guards back into the trap after the alarm is deactivated (p). Once again, all seven prisoners safely escape. This demonstrates our system's ability to react and adapt to static changes in the environment by generating alternative narrative plans.

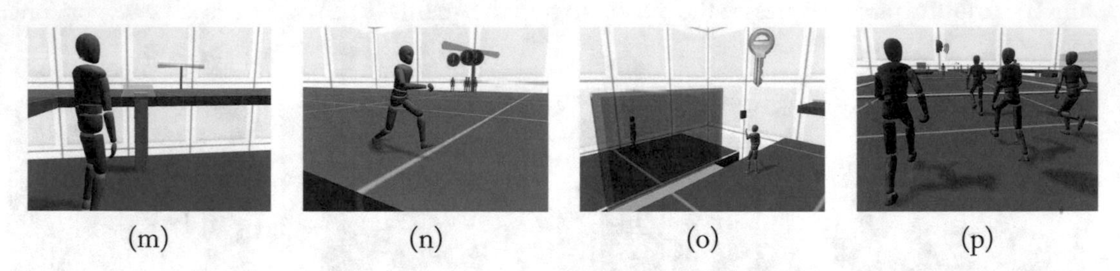

(m) (n) (o) (p)

Figure 22.7: Adaptations to the narrative plan resulting from an environment change.

22.4.5 REACTING TO USER INTERVENTION

Our system can perform rapid re-planning when faced with an invalided plan. We demonstrate two scenarios in which a user interacts with the system while a plan is currently active. In the first scenario, the user manually opens two doors in the north wing of the complex (near (4) in Fig. 22.5, light blue and purple). This prevents the prisoners in the north wing from having to open those doors by pressing their associated buttons. As soon as either door is opened, the planner detects the change, invalidates the active plan, and immediately produces a new plan from the current state of the world. Where the original plan instructed the prisoners to open those doors after distracting the guard blocking them, the new adapted plan allows the north wing prisoners to bypass that step entirely.

In a second scenario, the user acts in direct opposition to the goal, illustrated in Fig. 22.8. While the narrative is taking place, the user moves the four guards (q) from the cage room (near (4) in Fig. 22.5) closer to the door (r) at the exit of the south cellblock (near (3), blue). This prevents the prisoners from using the diversion tactic they employed in the original plans to trap the guards in the cage. To adapt to this change, the planner produces a new narrative where the prisoners use the alarm to lure the guards back out of the cage room (s), and intercept them with the cage as they pass over it (t). This showcases our system's ability to continue to progress the narrative towards an authored goal despite confounding intervention, intentional or otherwise, by a user.

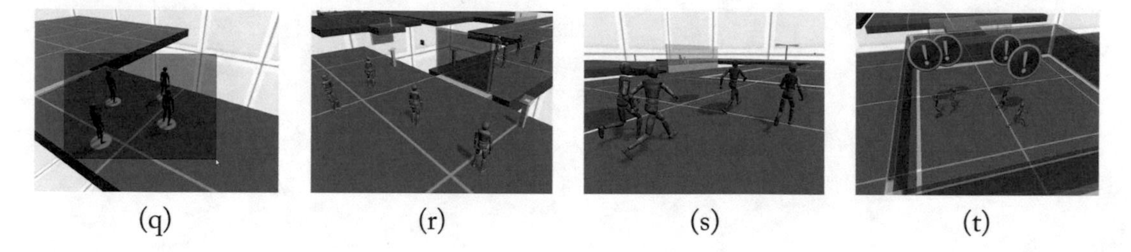

(q) (r) (s) (t)

Figure 22.8: The user selects and moves four guards (q, r), resulting in a real-time narrative plan adaptation (s, t).

CHAPTER 23

Conclusion

ADAPT is a modular, flexible platform which provides a comprehensive feature set for animation, navigation, and behavior tools needed for end-to-end simulation development. By allowing a user to independently incorporate a new animation choreographer or steering system, and make those components immediately accessible to the behavior level without modifying other existing systems, characters can very easily be expanded with new capabilities and functionality. ADAPT is released as an open-source project to allow users to tailor the system to fit their own personal needs, to rapidly iterate on experimental designs, and to compare their results against other established techniques.

ADAPT is designed to be modular and currently includes many of the steering, navigation, and behavior systems described in this book. Researchers and developers may additionally implement their own modules and easily interface with ADAPT depending on the needs of the application. The library, assets, and documentation are available at `http://cg.cis.upenn.edu/ADAPT`.

Using ADAPT, we have developed a framework for creating real-time interactive narratives using the event-centric authoring paradigm. Our method allows us to design complex and intricate multi-character interactions with a high degree of control fidelity in order to synchronize cooperative and competitive character behaviors. Planning in the space of events rather than in individual character action spaces allows characters to exhibit a large repertoire of individual capabilities without causing combinatorial growth in the planner's branching factor. This produces long, cohesive narratives at interactive rates. Event-centric authoring facilitates the synthesis of narratives with complex multi-actor coordination without the need for planning in the action space of individual agents. We advocate a hybrid system that balances event-centric coordination and agent-centric decision-making to synthesise narrative-driven interactive virtual worlds where autonomous avatars are able to participate in complex stories.

Our system imposes a considerable authoring burden which needs to be addressed before authoring digital stories can be truly made accessible to end users. Currently, events must be manually authored with their behavior, preconditions, postconditions, and cost. While feasible for a small number of events, this can quickly become prohibitive for larger, more interesting virtual worlds. Complex events can also grow unpredictable in nature, making it difficult to manually extrapolate their pre- and post conditions. To counteract the authoring burden for event design, we are actively exploring ways to partially automate the event authoring process. This revolves around a system that understands the nature of certain objects in the world, and the ramifications of interacting with them. Another exciting avenue of future exploration is to automatically synthesise interactive narratives [121, 135, 244]. This section focuses on the high-level aspects of multi-character coordination and behavior synthesis. The development of low-level movement capabilities for autonomous virtual humans, and in particular semantic annotation of motion databases [120] for motion synthesis is complementary to this work.

CHAPTER 24

Epilogue

This book provides a holistic bottom-to-top overview of several essential components that are needed in order to create realistic virtual crowds. These include footstep-level navigation and locomotion, machine learned steering, efficient discrete representations of the environment for navigation, sensory inputs and reactive control, individualized differences and personalities for crowd groups, and high-level planning and narrative storytelling. Many of the contributions described in this book can be found in open-source software implementation such as ADAPT [245] and SteerSuite [250]. Given the broad range and scope of software systems involved, the creation of a single unified system would itself be a major undertaking that would require additional resources beyond those that have been available. Unification would require the careful integration of the many disparate components developed by multiple independent research groups. Our intention, therefore, is to espouse an expansive view of virtual agent simulation with the hope that commercial products, such as games or virtual reality systems, will have the economic and marketing incentives to improve user experiences in interacting with groups of realistically animated *human-like* autonomous agents.

There are several related contributions that have not been included in this volume. These include the use of alibi generation to generate perceptually plausible location goals for members of a crowd who are being observed for a period of time [269], and the extensive work in perceptually guided crowd simulation [181, 182]. We have also left out techniques that are used in commercial games such as Grand Theft Auto, Assassin's Creed, and the Sims, which use similar principles and techniques, while enforcing strict robustness and performance constraints. We have also left out the detailed computer graphics techniques for both CPUs and GPUs that are needed to support large-scale crowds and highly realistic displays. A useful resource is the book by Thalmann and Musse [272]. Readers may also refer to related resources on navigation and real-time planning for interactive virtual worlds [114].

There is also a growing need to analyze and evaluate crowd simulation techniques in their ability to match real crowd datasets [80], to detect anomalies [22, 127], and to meet various quality and performance criteria [133, 134, 251, 255]. These quantitative metrics can also be used to automatically optimize the behavior of simulated crowds to satisfy user-defined objectives [13, 16] and environments [12, 14, 17].

Finally, we did not describe in this volume how these virtual crowd simulations can apply to real-world problem domains, such as large-scale crowd evacuation, stadium entry and egress, police crowd management, game environments, or movie special effects. Our research groups, as well as many others, have utilized advances in crowd simulation to address real pedestrian movement planning and evacuation safety problems. These remain challenging proving grounds for applying the techniques described here to real problems in agent scalability, domain knowledge generation, user narrative authoring, and validated animated experiences.

Bibliography

[1] Abaci, T. and Thalmann, D. (2005). Planning with smart objects. In *Intl. Conf. in Central Europe on Computer Graphics, Visualization and Computer Vision*, WSCG'05, pages 25–28. 178

[2] Ahn, J., Wang, N., Thalmann, D., and Boulic, R. (2012). Within-crowd immersive evaluation of collision avoidance behaviors. In *ACM SIGGRAPH Intl. Conf. on Virtual-Reality Continuum and Its Applications in Industry*, VRCAI '12, pages 231–238, New York, NY, USA. ACM. DOI: 10.1145/2407516.2407573. 8

[3] Allbeck, J. M. (2010). CAROSA: A tool for authoring NPCs. In *Motion in Games*, Lecture Notes in Computer Science, pages 182–193, Berlin, Heidelberg. Springer-Verlag. DOI: 10.1007/978-3-642-16958-8_18. 168

[4] Anderson, J. (2001). *How Can the Human Mind Exist in the Physical Universe?* Oxford University Press. 146

[5] Arikan, O., Chenney, S., and Forsyth, D. A. (2001). Efficient multi-agent path planning. In *Eurographics Workshop on Computer Animation and Simulation*, SCA"01, pages 151–162, New York, NY, USA. Springer-Verlag New York. DOI: 10.1007/978-3-7091-6240-8_14. 55

[6] Baddeley, A., Eysenck, M. W., and Anderson, M. C. (2012). *Memory*. Psychology Press, Taylor and Francis Inc. 151, 152

[7] Balint, J. T. and Allbeck, J. M. (2013a). MacGyver virtual agents: Using ontologies and hierarchies for resourceful virtual human decision-making. In *Intl. Conf. on Autonomous Agents and Multiagent Systems*, AAMAS '13, pages 1153–1154, Richland, SC. Intl. Foundation for Autonomous Agents and Multiagent Systems. 95

[8] Balint, J. T. and Allbeck, J. M. (2013b). What's going on? multi-sense attention for virtual agents. In *Intelligent Virtual Agents*, IVA'13, pages 349–357. DOI: 10.1007/978-3-642-40415-3_31. 128

[9] Balint, T. and Allbeck, J. M. (2014). Is that how everyone really feels? emotional contagion with masking for virtual crowds. In Bickmore, T., Marsella, S., and Sidner, C., editors, *Intelligent Virtual Agents*, volume 8637 of *Lecture Notes in Computer Science*, pages 26–35. Springer Intl. Publishing. DOI: 10.1007/978-3-319-09767-1. 173

[10] Bandi, S. and Thalmann, D. (1998). Space discretization for efficient human navigation. *Computer Graphics Forum*, 17(3):195–206. DOI: 10.1111/1467-8659.00267. 56

[11] Bee, M. A. and Micheyl, C. (2008). The cocktail party problem: What is it? how can it be solved? and why should animal behaviorists study it? *Journal of Comparative Psychology*, 122(3):235–251. DOI: 10.1037/0735-7036.122.3.235. 110

[12] Berseth, G., Haworth, M. B., Kapadia, M., and Faloutsos, P. (2014a). Characterizing and optimizing game level difficulty. In *Motion in Games*, pages 153–160, New York, NY, USA. ACM. DOI: 10.1145/2668084.2668100. 217

[13] Berseth, G., Kapadia, M., and Faloutsos, P. (2013a). Automated parameter tuning for steering algorithms. In *ACM SIGGRAPH/Eurographics Symp. on Computer Animation, Poster Proc.*, SCA '13, pages 115–124, New York, NY, USA. ACM. 217

[14] Berseth, G., Kapadia, M., and Faloutsos, P. (2013b). Steerplex: Estimating scenario complexity for simulated crowds. In *Motion on Games*, MIG '13, pages 45:67–45:76, New York, NY, USA. ACM. DOI: 10.1145/2522628.2522650. 217

[15] Berseth, G., Kapadia, M., and Faloutsos, P. (2015a). Robust space-time footsteps for agent-based steering. *Computer Animation and Virtual Worlds*. 15

[16] Berseth, G., Kapadia, M., Haworth, B., and Faloutsos, P. (2014b). SteerFit: Automated Parameter Fitting for Steering Algorithms. In *ACM SIGGRAPH/Eurographics Symp. on Computer Animation*, SCA '14, New York, NY, USA. ACM. DOI: 10.2312/sca.20141129. 217

[17] Berseth, G., Usman, M., Haworth, B., Kapadia, M., and Faloutsos, P. (2015b). Environment optimization for crowd evacuation. *Computer Animation and Virtual Worlds*, 26(3-4):377–386. DOI: 10.1002/cav.1652. 217

[18] Biddle, B. J. (1979). *Role Theory: Concepts and Research*. Krieger. 166, 167, 170

[19] Biknevicius, A. and Reilly, S. (2006). Correlation of symmetrical gaits and whole body mechanics: debunking myths in locomotor biodynamics. *Journal of Experimental Zoology Part A: Comparative Experimental Biology*, 305A(11):923–934. 18

[20] Bindiganavale, R., Schuler, W., Allbeck, J. M., Badler, N. I., Joshi, A. K., and Palmer, M. (2000). Dynamically altering agent behaviors using natural language instructions. In *Intl. Conf. on Autonomous Agents*, AGENTS '00, pages 293–300, New York, NY, USA. ACM. DOI: 10.1145/336595.337503. 95, 137

[21] Bird, S., Klein, E., and Loper, E. (2009). *Natural Language Processing with Python*. O'Reilly Media. 133

[22] Boatright, C. D., Kapadia, M., and Badler, N. I. (2012). Pedestrian anomaly detection using context-sensitive crowd simulation. In *Intl. Workshop on Pattern Recognition and Crowd Analysis*. 217

[23] Boatright, C. D., Kapadia, M., Shapira, J. M., and Badler, N. I. (2013). Context-sensitive data-driven crowd simulation. In *ACM SIGGRAPH Intl. Conf. on Virtual Reality Continuum and Its Applications in Industry*, VRCAI '13, pages 51–56, New York, NY, USA. ACM. DOI: 10.1145/2534329.2534332. 38

[24] Boatright, C. D., Kapadia, M., Shapira, J. M., and Badler, N. I. (2014). Generating a multiplicity of policies for agent steering in crowd simulation. *Computer Animation and Virtual Worlds*, pages n/a–n/a. DOI: 10.1002/cav.1572. 38

[25] Bonebright, T. L. (2001). Perceptual structure of everyday sounds: A multidimensional scaling approach. In *Intl. Conf. on Auditory Display*, pages 73–78. 101

[26] Botea, A., Müller, M., and Schaeffer, J. (2004). Near optimal hierarchical path-finding. *Journal of Game Development*, 1:7–28. 58

[27] Brom, C., Peskova, K., and Lukavsky, J. (2007). What does your actor remember? towards characters with a full episodic memory. In Cavazza, M. and Donikian, S., editors, *Virtual Storytelling. Using Virtual Reality Technologies for Storytelling*, volume 4871 of *Lecture Notes in Computer Science*, pages 89–101. Springer Berlin Heidelberg. DOI: 10.1007/978-3-540-77039-8. 145

[28] Bulitko, V., Sturtevant, N., Lu, J., and Yau, T. (2007). Graph abstraction in real-time heuristic search. *Journal of Artificial Intelliegence Research*, 30(1):51–100. DOI: 10.1613/jair.2293. 58

[29] Burke, R., Isla, D., Downie, M., Ivanov, Y., and Blumberg, B. (2001). Creature Smarts: The Art and Architecture of a Virtual Brain. 145, 151

[30] Canosa, R. L. (2005). Modeling selective perception of complex, natural scenes. *Intl. Journal on Artificial Intelligence Tools*, pages 233–260. DOI: 10.1142/S0218213005002089. 93

[31] Cavazza, M., Charles, F., and Mead, S. J. (2002). Character-based interactive storytelling. *IEEE Intelligent Systems*, 17(4):17–24. DOI: 10.1109/MIS.2002.1024747. 178

[32] Cavazza, M., Hartley, S., Lugrin, J.-L., and Le Bras, M. (2004). Qualitative physics in virtual environments. In *Intl. Conf. on Intelligent User Interfaces*, IUI '04, pages 54–61, New York, NY, USA. ACM. DOI: 10.1145/964442.964454. 95

[33] Cha, M., Cho, K., and Um, K. (2009). Design of memory architecture for autonomous virtual characters using visual attention and quad-graph. In *Intl. Conf. on Interaction Sciences:*

Information Technology, Culture and Human, ICIS '09, pages 691–696, New York, NY, USA. ACM. DOI: 10.1145/1655925.1656050. 94, 146

[34] Chai, J. and Hodgins, J. K. (2007). Constraint-based motion optimization using a statistical dynamic model. *ACM Trans. on Graphics*, 26(3). DOI: 10.1145/1276377.1276387. 9

[35] Choi, M. G., Kim, M., Hyun, K. L., and Lee, J. (2011). Deformable motion: Squeezing into cluttered environments. *Computer Graphics Forum*, 30(2):445–453. DOI: 10.1111/j.1467-8659.2011.01889.x. 58

[36] Choi, M. G., Lee, J., and Shin, S. Y. (2003). Planning biped locomotion using motion capture data and probabilistic roadmaps. *ACM Trans. on Graphics*, 22(2):182–203. DOI: 10.1145/636886.636889. 9, 58

[37] Chopra-Khullar, S. and Badler, N. I. (1999). Where to look? automating attending behaviors of virtual human characters. In *Intl. Conf. on Autonomous Agents*, AGENTS '99, pages 16–23, New York, NY, USA. ACM. DOI: 10.1023/A:1010010528443. 94

[38] Chung, S.-k. and Hahn, J. K. (1999). Animation of human walking in virtual environments. In *Computer Animation*, CA '99, pages 4–, Washington, DC, USA. IEEE Computer Society. DOI: 10.1109/CA.1999.781194. 14

[39] Coros, S., Beaudoin, P., Yin, K. K., and van de Panne, M. (2008). Synthesis of constrained walking skills. *ACM Trans. on Graphics*, 27(5):1–9. DOI: 10.1145/1409060.1409066. 14

[40] Cowling, M. and Sitte, R. (2003). Comparison of techniques for environmental sound recognition. *Pattern Recognition Letters*, 24(15):2895–2907. DOI: 10.1016/S0167-8655(03)00147-8. 141

[41] Curtis, S., Snape, J., and Manocha, D. (2012). Way portals: efficient multi-agent navigation with line-segment goals. In *ACM SIGGRAPH Symp. on Interactive 3D Graphics and Games*, I3D '12, pages 15–22, New York, NY, USA. ACM. DOI: 10.1145/2159616.2159619. 56

[42] Cycorp, Inc. (2007). OpenCyc Platform. `http://www.cyc.com/`. 135

[43] Deutsch, T., Gruber, A., Lang, R., and Velik, R. (2008). Episodic memory for autonomous agents. In *Human System Interactions*, pages 621–626. DOI: 10.1109/HSI.2008.4581512. 146

[44] Dias, J. a., Ho, W. C., Vogt, T., Beeckman, N., Paiva, A., and André, E. (2007). I know what i did last summer: Autobiographic memory in synthetic characters. In *Intl. Conf. on Affective Computing and Intelligent Interaction*, ACII '07, pages 606–617, Berlin, Heidelberg. Springer-Verlag. DOI: 10.1007/978-3-540-74889-2_53. 145

[45] Dictionary, W. C. (1991). 166

[46] Dodd, W. (2005). The design of procedural, semantic, and episodic memory systems for a cognitive robot. Master's thesis, Vanderbilt University. 146

[47] Donikian, S. (2009). Golaem. `http://www.golaem.com/`. 14

[48] Douglas, D. H. and Peucker, T. K. (1973). Algorithms for the reduction of the number of points required to represent a digitized line or its caricature. *Cartographica: The Intl. Journal for Geographic Information and Geovisualization*, 10(2):112–122. DOI: 10.3138/FM57-6770-U75U-7727. 66, 68

[49] Douville, B., Levison, L., and Badler, N. I. (1996). Task level object grasping for simulated agents. *Presence: Teleoperators and Virtual Environments*, 5(4):416–430. 95, 117, 132

[50] Durupinar, F., Allbeck, J., Pelechano, N., and Badler, N. (2008). Creating crowd variation with the ocean personality model. In *Intl. Joint Conf. on Autonomous Agents and Multiagent Systems*, AAMAS'08, pages 1217–1220, Richland, SC. Intl. Foundation for Autonomous Agents and Multiagent Systems. DOI: 10.1145/1402821.1402835. 7

[51] Durupinar, F., Pelechano, N., Allbeck, J., Güdükbay, U., and Badler, N. I. (2011). How the Ocean personality model affects the perception of crowds. *IEEE Computer Graphics and Applications*, 31(3):22–31. DOI: 10.1109/MCG.2009.105. 147, 161

[52] Ebbinghaus, H. (1964). *Memory: A Contribution to Experimental Psychology*. Dover. 152

[53] Eckert, K., Niepert, M., Niemann, C., Buckner, C., Allen, C., and Stuckenschmidt, H. (2010). Crowdsourcing the assembly of concept hierarchies. In *Joint Conf. on Digital Libraries*, JCDL '10, pages 139–148, New York, NY, USA. ACM. DOI: 10.1145/1816123.1816143. 132

[54] Egges, A. and van Basten, B. J. H. (2010). One Step at a Time: Animating Virtual Characters Based on Foot Placement. *The Visual Computer*, 26(6-8):497–503. DOI: 10.1007/s00371-010-0443-0. 9, 23

[55] Eisemann, E. and Décoret, X. (2006). Fast scene voxelization and applications. In *ACM Symp. on Interactive 3D Graphics*, I3D'06, pages 71–78. DOI: 10.1145/1111411.1111424. 60

[56] Ellenson, A. (1982). *Human Relations*. Prentice Hall College Div., 2 edition. 166

[57] Erol, K., Hendler, J., and Nau, D. S. (1994). Htn planning: Complexity and expressivity. In *National Conf. on Artificial Intelligence*, volume 2 of *AAAI'94*, pages 1123–1128, Menlo Park, CA, USA. American Assoc. for Artificial Intelligence. 178

[58] Erra, U., Frola, B., and Scarano, V. (2010). BehaveRT: a GPU-based library for autonomous characters. In *Motion in Games*, Lecture Notes in Computer Science, pages 194–205, Berlin, Heidelberg. Springer-Verlag. DOI: 10.1007/978-3-642-16958-8_19. 177

[59] Fan, J., Barker, K., Porter, B., and Clark, P. (2001). Representing roles and purpose. In *Intl. Conf. on Knowledge Capture*, K-CAP, pages 38–43. ACM. DOI: 10.1145/500737.500747. 169

[60] Felis, M. L. and Mombaur, K. (2012). Using Optimal Control Methods to Generate Human Walking Motions. In *Motion in Games*, Lecture Notes in Computer Science, pages 197–207, Berlin, Heidelberg. Springer-Verlag. DOI: 10.1007/978-3-642-34710-8_19. 9

[61] Feng, A., Huang, Y., Xu, Y., and Shapiro, A. (2012). Automating the transfer of a generic set of behaviors onto a virtual character. In Kallmann, M. and Bekris, K., editors, *Motion in Games*, volume 7660 of *Lecture Notes in Computer Science*, pages 134–145. Springer-Verlag, Berlin, Heidelberg. 127

[62] Fernandez, J., Canovas, L., and Pelegrin, B. (2000). Algorithms for the decomposition of a polygon into convex polygons. *European Journal of Operational Research*, 121(2):330–342. DOI: 10.1016/S0377-2217(99)00033-8. 68

[63] Feurtey, F. (2000). Simulating the Collision Avoidance Behavior of Pedestrians. Master's thesis, University of Tokyo, Department of Electronic Engineering. 8

[64] Fikes, R. E. and Nilsson, N. J. (1971). STRIPS: a new approach to the application of theorem proving to problem solving. pages 608–620. 126, 178, 202

[65] Fleischman, M. and Roy, D. (2007). Representing intentions in a cognitive model of language acquisition: Effects of phrase structure on situated verb learning. In *American Assoc. for Artificial Intelligence*, pages 7–12. 177

[66] Fraichard, T. (1999). Trajectory planning in a dynamic workspace: A 'state-time space' approach. *Advanced Robotics*, 13(1):75–94. DOI: 10.1163/156855399X01017. 58

[67] Franklin, S. and Patterson, Jr., F. G. (2006). The lida architecture: Adding new modes of learning to an intelligent, autonomous, software agent. In *Integrated Design and Process Technology*. 146

[68] Funge, J., Tu, X., and Terzopoulos, D. (1999). Cognitive modeling: Knowledge, reasoning and planning for intelligent characters. In *ACM SIGGRAPH Conf. on Computer Graphics and Interactive Techniques*, SIGGRAPH '99, pages 29–38, New York, NY, USA. ACM Press/Addison-Wesley. DOI: 10.1145/311535.311538. 178

[69] Garcia, F. M., Kapadia, M., and Badler, N. I. (2014). Gpu-based dynamic search on adaptive resolution grids. In *IEEE Intl. Conf. on Robotics and Automation*, ICRA'14, pages 1631–1638. DOI: 10.1109/ICRA.2014.6907070. 56

[70] Geraerts, R. (2010). Explicit Corridor Map. `http://www.staff.science.uu.nl/~gerae101/motion_planning/cm/index.html`. 57

[71] Gibson, J. J. (1986). *The Ecological Approach to Visual Perception*. Routledge. 94

[72] Girard, M. (1987). Interactive design of 3d computer-animated legged animal motion. *IEEE Computer Graphics and Applications*, 7(6):39–51. DOI: 10.1109/MCG.1987.276895. 14

[73] Girard, M. and Maciejewski, A. A. (1985). Computational modeling for the computer animation of legged figures. 19(3):263–270. DOI: 10.1145/325165.325244. 9

[74] Glardon, P., Boulic, R., and Thalmann, D. (2006). Robust on-line adaptive footplant detection and enforcement for locomotion. *The Visual Computer*, 22(3):194–209. DOI: 10.1007/s00371-006-0376-9. 9

[75] Gochev, K., Cohen, B. J., Butzke, J., Safonova, A., and Likhachev, M. (2011). Path planning with adaptive dimensionality. In *Symp. on Combinatorial Search*, SCS'11. 58

[76] Goldberg, L. R. (1990). An alternative "Description of Personality": The big-five factor structure. *Journal of Personality and Social Psychology*, 59:1216–1229. DOI: 10.1037/0022-3514.59.6.1216. 159

[77] Gomes, P. F., Martinho, C., and Paiva, A. (2011). I've been here before!: Location and appraisal in memory retrieval. In *Intl. Conf. on Autonomous Agents and Multiagent Systems*, volume 3 of *AAMAS '11*, pages 1039–1046, Richland, SC. Intl. Foundation for Autonomous Agents and Multiagent Systems. 145

[78] Gu, E. and Badler, N. I. (2006). Visual attention and eye gaze during multiparty conversations with distractions. In *Intelligent Virtual Agents*, IVA'06, pages 193–204, Berlin, Heidelberg. Springer-Verlag. DOI: 10.1007/11821830_16. 94

[79] Guy, S. J., Chhugani, J., Kim, C., Satish, N., Lin, M., Manocha, D., and Dubey, P. (SCA'09). Clearpath: highly parallel collision avoidance for multi-agent simulation. In *ACM SIGGRAPH/Eurographics Symp. on Computer Animation*, pages 177–187. 8

[80] Guy, S. J., van den Berg, J., Liu, W., Lau, R., Lin, M. C., and Manocha, D. (2012). A statistical similarity measure for aggregate crowd dynamics. *ACM Trans. on Graphics*, 31(6):190:1–190:11. DOI: 10.1145/2366145.2366209. 217

[81] Gygi, B., Kidd, G. R., and Watson, C. S. (2007). Similarity and categorization of environmental sounds. *Attention, Perception, & Psychophysics*, 69(6):839–855. DOI: 10.3758/BF03193921. 94, 98, 100, 111

[82] Hale, D. H. and Youngblood, G. M. (2009). Full 3D spatial decomposition for the generation of navigation meshes. In Darken, C. and Youngblood, G. M., editors, *Artificial Intelligence and Interactive Digital Entertainment*, AIIDE'09. The AAAI Press. 56

[83] Hale, D. H., Youngblood, G. M., and Dixit, P. N. (2008). Automatically-generated convex region decomposition for real-time spatial agent navigation in virtual worlds. In *Artificial Intelligence and Interactive Digital Entertainment Conf.*, AIIDE'08. 56

[84] Hall, E. T. (1966). *The Hidden Dimension*. Anchor Books. DOI: 10.2307/1572461. 162

[85] Hart, P. E., Nilsson, N. J., and Raphael, B. (1972). Correction to "A Formal Basis for the Heuristic Determination of Minimum Cost Paths". *SIGART Bulletin*, (37):28–29. DOI: 10.1145/1056777.1056779. 178

[86] Haumont, D., Debeir, O., and Sillion, F. X. (2003). Volumetric cell-and-portal generation. *Computer Graphics Forum*, 22(3):303–312. DOI: 10.1111/1467-8659.00677. 57

[87] Helbing, D., Farkas, I., and Vicsek, T. (2000). Simulating dynamical features of escape panic. *Nature*, 407:487–490. DOI: 10.1038/35035023. 23

[88] Helbing, D. and Molnár, P. (1995). Social force model for pedestrian dynamics. *Physical Review E*, 51(5):4282–4286. DOI: 10.1103/PhysRevE.51.4282. 7

[89] Herrero, P. and de Antonio, A. (2003). Introducing human-like hearing perception in intelligent virtual agents. In *Intl. Joint Conf. on Autonomous Agents and Multiagent Systems*, AAMAS'03, pages 733–740. ACM. DOI: 10.1145/860575.860693. 93

[90] Herrero, P., Greenhalgh, C., and de Antonio, A. (2005). Modelling the sensory abilities of intelligent virtual agents. *Autonomous Agents and Multiagent Systems*, 11(3):361–385. DOI: 10.1007/s10458-005-2921-8. 93, 94, 120

[91] Hill, Jr., R. W. (2000). Perceptual attention in virtual humans: Towards realistic and believable gaze behaviors. In *AAAI Fall Symp. on Simulating Human Agents*, pages 46–52. 117, 118

[92] Hill, Jr., R. W., Kim, Y., and Gratch, J. (2002). Anticipating where to look: predicting the movements of mobile agents in complex terrain. In *Intl. Conf. on Autonomous Agents and Multiagent Systems*, volume 2 of *AAMAS*, pages 821–827, Bologna, Italy. DOI: 10.1145/544862.544935. 122

[93] Ho, W. C., Dias, J. a., Figueiredo, R., and Paiva, A. (2007). Agents that remember can tell stories: Integrating autobiographic memory into emotional agents. In *Intl. Joint Conf. on Autonomous Agents and Multiagent Systems*, AAMAS '07, pages 10:1–10:3, New York, NY, USA. ACM. DOI: 10.1145/1329125.1329138. 145

[94] Hochberg, J. and McAlister, E. (1953). A quantitative approach to figural goodness. *Journal of Experimental Psychology*, 46(5):361–364. DOI: 10.1037/h0055809. 118

[95] Hoff III, K., Culver, T., Keyser, J., Lin, M. C., and Manocha, D. (2000). Interactive motion planning using hardware-accelerated computation of generalized voronoi diagrams. In *IEEE Intl. Conf. on Robotics and Automation*, volume 3 of *ICRA'00*, pages 2931–2937. DOI: 10.1109/ROBOT.2000.846473. 58

[96] Holte, R. C., Grajkowski, J., and Tanner, B. (2005). Hierarchical heuristic search revisited. In *Abstraction, Reformulation and Approximation*, volume 3607 of *LNCS*, pages 121–133. Springer Berlin Heidelberg. DOI: 10.1007/11527862_9. 58

[97] Holte, R. C., Perez, M. B., Zimmer, R. M., and MacDonald, A. J. (1996). Hierarchical A*: Searching abstraction hierarchies efficiently. In *National conference on Artificial intelligence*, AAAI'96, pages 530–535. AAAI Press. 58

[98] Hory, C., Martin, N., and Chehikian, A. (2002). Spectrogram segmentation by means of statistical features for non-stationary signal interpretation. *IEEE Trans. on Signal Processing*, 50(12):2915–2925. DOI: 10.1109/TSP.2002.805489. 100

[99] Hsu, D., Kindel, R., Latombe, J.-C., and Rock, S. (2002). Randomized kinodynamic motion planning with moving obstacles. *The Intl. Journal of Robotics Research*, 21(3):233–255. DOI: 10.1177/027836402320556421. 58

[100] Huang, T., Kapadia, M., Badler, N. I., and Kallmann, M. (2014). Path planning for coherent and persistent groups. In *IEEE Intl. Conf. on Robotics and Automation*, ICRA '14, pages 1652–1659. IEEE. DOI: 10.1109/ICRA.2014.6907073. 51, 57

[101] Ickes, W. and Knowles, E. S. (1982). *Personality, Roles, and Social Behavior*. Springer. DOI: 10.1007/978-1-4613-9469-3. 167

[102] Isla, D. (2005). Handling complexity in the Halo 2 AI. In *Game Developers Conf.* 194

[103] James, D. L., Barbič, J., and Pai, D. K. (2006). Precomputed acoustic transfer: output-sensitive, accurate sound generation for geometrically complex vibration sources. *ACM Trans. on Graphics*, 25(3):987–995. DOI: 10.1145/1141911.1141983. 94

[104] Jhala, A., Rawls, C., Munilla, S., and Young, R. M. (2008). Longboard: A sketch based intelligent storyboarding tool for creating machinima. In *FLAIRS Conf.*, pages 386–390. AAAI Press. 178

[105] Johansen, R. S. (2009). Automated semi-procedural animation for character locomotion. Master's thesis, Aarhus University. 9, 24, 25, 26, 30, 189

[106] Jordao, K., Pettré, J., Christie, M., and Cani, M.-P. (2014). Crowd Sculpting: A space-time sculpting method for populating virtual environments. *Computer Graphics Forum*, 33(2). DOI: 10.1111/cgf.12316. 177

[107] Jorgensen, C.-J. and Lamarche, F. (2011). From geometry to spatial reasoning: automatic structuring of 3d virtual environments. In *Motion in Games*, Lecture Notes in Computer Science, pages 353–364, Berlin, Heidelberg. Springer-Verlag. DOI: 10.1007/978-3-642-25090-3_30. 57

[108] Kagawa, Y., Tsuchiya, T., Fujii, B., and Fujioka, K. (1998). Discrete Huygens' model approach to sound wave propagation. *Journal of Sound and Vibration*, 218(3):419–444. DOI: 10.1006/jsvi.1998.1861. 98

[109] Kajita, S., Kanehiro, F., Kaneko, K., Yokoi, K., and Hirukawa, H. (2001). The 3d linear inverted pendulum mode: a simple modeling for a biped walking pattern generation. In *IEEE/RSJ Intl. Conf. on Intelligent Robots and Systems*, volume 1 of *IROS'01*, pages 239–246. DOI: 10.1109/IROS.2001.973365. 15

[110] Kalicinski, M. (2009). RapidXML. `http://rapidxml.sourceforge.net/`. 137

[111] Kallmann, M. (2005). Path planning in triangulations. In *IJCAI Workshop on Reasoning, Representation, and Learning in Computer Games*, Edinburgh, Scotland. 56

[112] Kallmann, M. (2010). Shortest paths with arbitrary clearance from navigation meshes. In *ACM SIGGRAPH/Eurographics Symp. on Computer Animation*, SCA '10, pages 159–168, Aire-la-Ville, Switzerland, Switzerland. Eurographics Assoc. DOI: 10.2312/SCA. 55, 56, 58, 76

[113] Kallmann, M., Bieri, H., and Thalmann, D. (2003). Fully dynamic constrained delaunay triangulations. In Brunnett, G., Hamann, B., Mueller, H., and Linsen, L., editors, *Geometric Modeling for Scientific Visualization*, pages 241–257. Springer-Verlag, Heidelberg, Germany. ISBN 3-540-40116-4. DOI: 10.1007/978-3-662-07443-5. 56

[114] Kallmann, M. and Kapadia, M. (2014). Navigation meshes and real-time dynamic planning for virtual worlds. In *ACM SIGGRAPH 2014 Course*, SIGGRAPH '14, pages 3:1–3:81, New York, NY, USA. ACM. DOI: 10.1145/2614028.2615399. 217

[115] Kallmann, M. and Thalmann, D. (1998). Modeling objects for interaction tasks. In *Eurographics Workshop on Animation and Simulation*, pages 73–86. DOI: 10.1007/978-3-7091-6375-7_6. 117, 119

[116] Kallmann, M. and Thalmann, D. (2002). Modeling behaviors of interactive objects for real time virtual environments. *Journal of Visual Languages and Computing*, 13:177–195. DOI: 10.1006/jvlc.2001.0229. 95, 132

[117] Kalogerakis, E., Christodoulakis, S., and Moumoutzis, N. (2006). Coupling ontologies with graphics content for knowledge driven visualization. In *IEEE Conf. on Virtual Reality*, VR '06, pages 43–50, Washington, DC, USA. IEEE Computer Society. DOI: 10.1109/VR.2006.41. 95

[118] Kapadia, M. and Badler, N. I. (2013). Navigation and steering for autonomous virtual humans. *Wiley Interdisciplinary Reviews: Cognitive Science*. DOI: 10.1002/wcs.1223. 177

[119] Kapadia, M., Beacco, A., Garcia, F., Reddy, V., Pelechano, N., and Badler, N. I. (2013a). Multi-domain real-time planning in dynamic environments. In *ACM SIGGRAPH/Eurographics Symp. on Computer Animation*, SCA '13, pages 115–124, New York, NY, USA. ACM. DOI: 10.1145/2485895.2485909. 76

[120] Kapadia, M., Chiang, I.-k., Thomas, T., Badler, N. I., and Kider, Jr., J. T. (2013b). Efficient motion retrieval in large motion databases. In *ACM SIGGRAPH Symp. on Interactive 3D Graphics and Games*, I3D '13, pages 19–28, New York, NY, USA. ACM. DOI: 10.1145/2448196.2448199. 216

[121] Kapadia, M., Falk, J., Zünd, F., Marti, M., Sumner, R. W., and Gross, M. (2015a). Computer-assisted authoring of interactive narratives. In *ACM Symp. on Interactive 3D Graphics and Games*, I3D '15, pages 85–92, New York, NY, USA. ACM. DOI: 10.1145/2699276.2699279. 178, 216

[122] Kapadia, M., Garcia, F., Boatright, C. D., and Badler, N. I. (2013c). Dynamic search on the GPU. In *IEEE/RSJ Intl. Conf. on Intelligent Robots and Systems*, IROS'13, pages 3332–3337. DOI: 10.1109/IROS.2013.6696830. 56

[123] Kapadia, M., Marshak, N., and Badler, N. I. (2014). Adapt: The agent development and prototyping testbed. *IEEE Trans. on Visualization and Computer Graphics*, 20(7):1035–1047. DOI: 10.1109/TVCG.2013.251. 179

[124] Kapadia, M., Ninomiya, K., Shoulson, A., Garcia, F., and Badler, N. (2013d). Constraint-aware navigation in dynamic environments. In *Motion on Games*, MIG '13, pages 89:111–89:120, New York, NY, USA. ACM. DOI: 10.1145/2522628.2522654. 58

[125] Kapadia, M., Shoulson, A., Boatright, C. D., Huang, P., Durupinar, F., and Badler, N. I. (2012a). What's next? the new era of autonomous virtual humans. In *Motion in Games*, Lecture Notes in Computer Science, pages 170–181, Berlin, Heidelberg. Springer-Verlag. DOI: 10.1007/978-3-642-34710-8_16. 177

[126] Kapadia, M., Shoulson, A., Durupinar, F., and Badler, N. I. (2013e). Authoring Multi-actor Behaviors in Crowds with Diverse Personalities. In Ali, S., Nishino, K., Manocha, D., and Shah, M., editors, *Modeling, Simulation and Visual Analysis of Crowds*, volume 11 of *The Intl. Series in Video Computing*, pages 147–180. Springer New York. DOI: 10.1007/978-1-4614-8483-7. 145, 177

[127] Kapadia, M., Singh, S., Allen, B., Reinman, G., and Faloutsos, P. (2009a). Steerbug: an interactive framework for specifying and detecting steering behaviors. In *ACM SIG-*

GRAPH/Eurographics Symp. on Computer Animation, SCA '09, pages 209–216, New York, NY, USA. ACM. DOI: 10.1145/1599470.1599497. 217

[128] Kapadia, M., Singh, S., Hewlett, W., and Faloutsos, P. (2009b). Egocentric affordance fields in pedestrian steering. In *ACM Symp. on Interactive 3D Graphics and Games*, I3D'09, pages 215–223, New York, NY, USA. ACM. DOI: 10.1145/1507149.1507185. 8

[129] Kapadia, M., Singh, S., Hewlett, W., Reinman, G., and Faloutsos, P. (2012b). Parallelized egocentric fields for autonomous navigation. *The Visual Computer*, 28(12):1209–1227. DOI: 10.1007/s00371-011-0669-5. 8

[130] Kapadia, M., Singh, S., Reinman, G., and Faloutsos, P. (2011a). Behavior authoring for crowd simulations. In *ACM SIGGRAPH Symp. on Interactive 3D Graphics and Games*, I3D '11, pages 199–199, New York, NY, USA. ACM. DOI: 10.1145/1944745.1944779. 178

[131] Kapadia, M., Singh, S., Reinman, G., and Faloutsos, P. (2011b). A behavior-authoring framework for multiactor simulations. *IEEE Computer Graphics and Applications*, 31(6):45 –55. DOI: 10.1109/MCG.2011.68. 114, 177, 178

[132] Kapadia, M., Singh, S., Reinman, G., and Faloutsos, P. (2011c). Multi-actor planning for directable simulations. In *Digital Media and Digital Content Management*, DMDCM '11, pages 111–116, Washington, DC, USA. IEEE Computer Society. DOI: 10.1109/DMDCM.2011.40. 177

[133] Kapadia, M., Wang, M., Reinman, G., and Faloutsos, P. (2011d). Improved benchmarking for steering algorithms. In *Motion in Games*, Lecture Notes in Computer Science, pages 266–277, Berlin, Heidelberg. Springer-Verlag. DOI: 10.1007/978-3-642-25090-3_23. 217

[134] Kapadia, M., Wang, M., Singh, S., Reinman, G., and Faloutsos, P. (2011e). Scenario space: characterizing coverage, quality, and failure of steering algorithms. In *ACM SIGGRAPH/Eurographics Symp. on Computer Animation*, SCA '11, pages 53–62, New York, NY, USA. ACM. 38, 84, 217

[135] Kapadia, M., Zünd, F., Falk, J., Marti, M., Sumner, R. W., and Gross, M. (2015b). Evaluating the authoring complexity of interactive narratives with interactive behaviour trees. In *Foundations of Digital Games*, FDG'15. 216

[136] Karamouzas, I., Heil, P., van Beek, P., and Overmars, M. H. (2009). A predictive collision avoidance model for pedestrian simulation. In *Motion in Games*, volume 5884 of *Lecture Notes in Computer Science*, pages 41–52, Berlin, Heidelberg. Springer-Verlag. DOI: 10.1007/978-3-642-10347-6_4. 8

[137] Karp, R. M. (2010). Reducibility among combinatorial problems. pages 219–241. 39

[138] Kim, J., Seol, Y., Kwon, T., and Lee, J. (2014). Interactive manipulation of large-scale crowd animation. *ACM Trans. on Graphics*, 33. DOI: 10.1145/2601097.2601170. 177

[139] Kim, M., Hwang, Y., Hyun, K., and Lee, J. (2012). Tiling motion patches. In *ACM SIGGRAPH/Eurographics Symp. on Computer Animation*, SCA'12, pages 117–126. DOI: 10.1109/TVCG.2013.80. 177

[140] Kim, M., Hyun, K., Kim, J., and Lee, J. (2009). Synchronized multi-character motion editing. *ACM Trans. on Graphics*, 28(3):79:1–79:9. DOI: 10.1145/1618452.1618507. 177

[141] Kim, Y., van Velsen, M., and Hill, Jr., R. W. (2005). Lecture notes in computer science. chapter Modeling Dynamic Perceptual Attention in Complex Virtual Environments, pages 266–277. Springer-Verlag, London, UK, UK. 93, 94

[142] Ko, H. and Badler, N. I. (1992). Straight line walking animation based on kinematic generalization that preserves the original characteristics, technical report. 14

[143] Ko, H. and Badler, N. I. (1996). Animating human locomotion with inverse dynamics. *IEEE Computer Graphics & Applications*, 16(2):50–59. DOI: 10.1109/38.486680. 9

[144] Kring, A. W., Champandard, A. J., and Samarin, N. (2010). DHPA* and SHPA*: Efficient Hierarchical Pathfinding in Dynamic and Static Game Worlds. In *Artificial Intelligence and Interactive Digital Entertainment*, AIIDE'10. The AAAI Press. 58

[145] Kristiansen, U. and Viggen, E. (2010). Computational methods in acoustics. Technical report, NTNU. 98, 105

[146] Kuffner, Jr., J. J. (1998). Goal-directed navigation for animated characters using real-time path planning and control. In *Workshop on Modelling and Motion Capture Techniques for Virtual Environments*, CAPTECH '98, pages 171–186, London, UK, UK. Springer-Verlag. DOI: 10.1007/3-540-49384-0_14. 56

[147] Kuffner, Jr., J. J. and Latombe, J.-C. (1999). Fast synthetic vision, memory, and learning models for virtual humans. In *Computer Animation*, pages 118–127. DOI: 10.1109/CA.1999.781205. 145

[148] Kuo, A. D. (2007). The six determinants of gait and the inverted pendulum analogy: A dynamic walking perspective. *Human Movement Science*, 26(4):617 – 656. European Workshop on Movement Science 2007. DOI: 10.1016/j.humov.2007.04.003. 18

[149] Kwon, T., Lee, K. H., Lee, J., and Takahashi, S. (2008). Group motion editing. *ACM Trans. on Graphics*, 27(3):80:1–80:8. DOI: 10.1145/1360612.1360679. 177

[150] Lacaze, A. (2002). Hierarchical planning algorithms. In *SPIE Intl. Symp. on Aerospace/Defense Sensing, Simulation, and Controls*. 58

[151] Laird, J. E. (2001). Using a computer game to develop advanced AI. *Computer*, 34(7):70–75. DOI: 10.1109/2.933506. 146

[152] Laird, J. E. (2012). *The Soar Cognitive Architecture*. The MIT Press. 146

[153] Lamarche, F. (2009). Topoplan: a topological path planner for real time human navigation under floor and ceiling constraints. *Computer Graphics Forum*, 28(2):649–658. DOI: 10.1111/j.1467-8659.2009.01405.x. 57

[154] Lamarche, F. and Donikian, S. (2004). Crowd of virtual humans: a new approach for real time navigation in complex and structured environments. *Computer Graphics Forum*, 23:509–518. DOI: 10.1111/j.1467-8659.2004.00782.x. 7

[155] Lau, M. and Kuffner, Jr., J. J. (2005). Behavior planning for character animation. In *ACM SIGGRAPH/Eurographics Symp. on Computer Animation*, SCA'05, pages 271–280. DOI: 10.1145/1073368.1073408. 58

[156] LaValle, S. M. (2006). *Planning Algorithms*. Cambridge University Press, Cambridge, U.K. Available at http://planning.cs.uiuc.edu/. DOI: 10.1017/CBO9780511546877. 14

[157] Lee, K. H., Choi, M. G., Hong, Q., and Lee, J. (2007). Group behavior from video: A data-driven approach to crowd simulation. In *ACM SIGGRAPH/Eurographics Symp. on Computer Animation*, SCA '07, pages 109–118, Aire-la-Ville, Switzerland, Switzerland. Eurographics Assoc. DOI: 10.1145/1272690.1272706. 8, 40

[158] Lee, K. H., Choi, M. G., and Lee, J. (2006). Motion patches: Building blocks for virtual environments annotated with motion data. *ACM Trans. on Graphics*, 25(3):898–906. DOI: 10.1145/1141911.1141972. 177

[159] Lerner, A., Chrysanthou, Y., and Lischinski, D. (2007). Crowds by example. *CGF*, 26(3):655–664. DOI: 10.1111/j.1467-8659.2007.01089.x. 8, 44

[160] Levine, S., Lee, Y., Koltun, V., and Popović, Z. (2011). Space-time planning with parameterized locomotion controllers. *ACM Trans. on Graphics*, 30:23:1–23:11. DOI: 10.1145/1966394.1966402. 58, 76

[161] Li, B. and Riedl, M. O. (2011). *Creating Customized Game Experiences by Leveraging Human Creative Effort: A Planning Approach*, volume 6525 of *Lecture Notes in Computer Science*, pages 99–116. Springer Berlin Heidelberg. 178

[162] Li, S.-X. and Loew, M. H. (1987). Adjacency detection using quadcodes. *Communications of the ACM*, 30(7):627–631. DOI: 10.1145/28569.28574. 109

[163] Li, W. and Allbeck, J. M. (2011). Populations with purpose. In *Motion in Games*, Lecture Notes in Computer Science, pages 132–143, Berlin, Heidelberg. Springer-Verlag. DOI: 10.1007/978-3-642-25090-3_12. 147, 169

[164] Li, W. and Allbeck, J. M. (2012a). The virtual apprentice. In Nakano, Y., Neff, M., Paiva, A., and Walker, M., editors, *Intelligent Virtual Agents*, volume 7502 of *Lecture Notes in Computer Science*, pages 15–27. Springer Berlin Heidelberg. DOI: 10.1007/978-3-642-33197-8. 145

[165] Li, W. and Allbeck, J. M. (2012b). The virtual apprentice. In *Intelligent Virtual Agents*, IVA'12, pages 15–27. DOI: 10.1007/978-3-642-33197-8_2. 177

[166] Li, W. and Allbeck, J. M. (2012c). Virtual humans: Evolving with common sense. In Kallmann, M. and Bekris, K., editors, *Motion in Games*, volume 7660 of *Lecture Notes in Computer Science*, pages 182–193. Springer Berlin Heidelberg, Berlin, Heidelberg. 119

[167] Likhachev, M., Ferguson, D. I., Gordon, G. J., Stentz, A., and Thrun, S. (2005). Anytime Dynamic A*: An anytime, replanning algorithm. In *Intl. Conf. on Automated Planning and Scheduling*, pages 262–271. 206

[168] Likhachev, M., Gordon, G. J., and Thrun, S. (2003). ARA*: Anytime A* with provable bounds on sub-optimality. In *Neural Information Processing Systems*. 58

[169] Lim, M. (2012). Memory models for intelligent social companions. In Zacarias, M. and de Oliveira, J., editors, *Human-Computer Interaction: The Agency Perspective*, volume 396 of *Studies in Computational Intelligence*, pages 241–262. Springer Berlin Heidelberg. DOI: 10.1007/978-3-642-25691-2. 146

[170] Lo, W.-Y. and Zwicker, M. (2008). Real-time planning for parameterized human motion. In *ACM SIGGRAPH/Eurographics Symp. on Computer Animation*, SCA'08, pages 29–38. DOI: 10.2312/SCA. 58

[171] Lopez, T., Lamarche, F., and Li, T.-Y. (2012). Space-time planning in changing environments : using dynamic objects for accessibility. *Computer Animation and Virtual Worlds*, 23(2):87–99. DOI: 10.1002/cav.1428. 58

[172] Loscos, C., Marchal, D., and Meyer, A. (2003). Intuitive crowd behaviour in dense urban environments using local laws. In *Theory and Practice of Computer Graphics*, TPCG '03, pages 122–, Washington, DC, USA. IEEE Computer Society. DOI: 10.1109/TPCG.2003.1206939. 7

[173] Loyall, A. B. (1997). *Believable Agents: Building Interactive Personalities*. PhD thesis, Carnegie Mellon University. 177

[174] Lugrin, J.-L. and Cavazza, M. (2007). Making sense of virtual environments: Action representation, grounding and common sense. In *Intl. Conf. on Intelligent User Interfaces*, IUI '07, pages 225–234, New York, NY, USA. ACM. DOI: 10.1145/1216295.1216336. 95

[175] Magerko, B., Laird, J. E., Assanie, M., Kerfoot, A., and Stokes, D. (2004). AI characters and directors for interactive computer games. In *Innovative Applications of Artificial Intelligence*, IAAI'04, pages 877–883. AAAI Press. 178

[176] Maslow, A. H. (1943). A theory of human motivation. *Psychological Review*, 50:370–396. DOI: 10.1037/h0054346. 167

[Massachusetts Institute of Technology] Massachusetts Institute of Technology. Conceptnet. `http://conceptnet5.media.mit.edu`. Online; accessed November 2014. 132

[178] Mateas, M. (2002). Interactive drama, art and artificial intelligence. Technical report, Carnegie Mellon University. 145

[179] Mateas, M. and Stern, A. (2003). Integrating plot, character and natural language processing in the interactive drama Façade. In *Technologies for Interactive Digital Storytelling and Entertainment*, volume 2. 177, 178

[180] Mcdermott, D., Ghallab, M., Howe, A., Knoblock, C., Ram, A., Veloso, M., Weld, D., and Wilkins, D. (1998). Pddl - the planning domain definition language. Technical Report TR-98-003, Yale Center for Computational Vision and Control,. 178

[181] McDonnell, R., Larkin, M., Dobbyn, S., Collins, S., and O'Sullivan, C. (2008). Clone attack! perception of crowd variety. *ACM Trans. on Graphics*, 27(3). DOI: 10.1145/1360612.1360625. 217

[182] McDonnell, R., Larkin, M., Hernández, B., Rudomín, I., and O'Sullivan, C. (2009). Eyecatching crowds: saliency based selective variation. *ACM Trans. on Graphics*, 28(3). DOI: 10.1145/1531326.1531361. 217

[183] McGinnies, E. (1994). *Perspectives on Social Behavior*. Gardner Press. 166, 168

[184] McGrenere, J. (2000). Affordances: Clarifying and evolving a concept. In *Graphics Interface*, GI'00, pages 179–186. 94

[185] Menou, E. (2001). Real-time character animation using multi-layered scripts and spacetime optimization. In *Intl. Conf. on Virtual Storytelling: Using Virtual Reality Technologies for Storytelling*, ICVS '01, pages 135–144, London, UK, UK. Springer-Verlag. DOI: 10.1007/3-540-45420-9_15. 177

[186] Merton, R. K. (1998). *Social Theory and Social Structure*. Free Press. 169

[187] Miller, G. A. (1956). The magical number seven, plus or minus two: some limits on our capacity for processing information. *Psychological Review*, 63(2):81–97. DOI: 10.1037/h0043158. 128, 151

[188] Minsky, M. (1986). *The Society of Mind*. Simon and Schuster. 151

[189] Mononen, M. (2009a). Recast navigation toolkit. http://code.google.com/p/recastnavigation/. 57

[190] Mononen, M. (2009b). Recast/Detour Navigation Library. http://code.google.com/p/recastnavigation/. 184

[191] Monzani, J.-S. and Thalmann, D. (2000). A sound propagation model for interagents communication. In *Intl. Conf. on Virtual Worlds*, VW '00, pages 135–146, London, UK, UK. Springer-Verlag. DOI: 10.1007/3-540-45016-5_13. 94

[192] Moulin, B. (1998). The social dimension of interactions in multiagent systems. *Agents and Multi-agent Systems, Formalisms, Methodologies, and Applications*, 1441/1998. DOI: 10.1007/BFb0055023. 169

[193] Moussaid, M., Helbing, D., and Theraulaz, G. (2011). How simple rules determine pedestrian behavior and crowd disasters. *Proc. of the National Academy of Sciences*, 108(17):6884–6888. DOI: 10.1073/pnas.1016507108. 7

[194] Narain, R., Golas, A., Curtis, S., and Lin, M. C. (2009). Aggregate dynamics for dense crowd simulation. *ACM Trans. on Graphics*, 28(5):1. DOI: 10.1145/1618452.1618468. 7

[195] Ninomiya, K., Kapadia, M., Shoulson, A., Garcia, F., and Badler, N. (2014). Planning approaches to constraint-aware navigation in dynamic environments. *Computer Animation and Virtual Worlds*, pages n/a–n/a. DOI: 10.1002/cav.1622. 58

[196] Normoyle, A., Liu, F., Kapadia, M., Badler, N. I., , and Joerg, S. (2013). The effect of posture and dynamics on the perception of emotion. In *ACM Symp. on Applied Perception*, SAP '13, New York, NY, USA. ACM. DOI: 10.1145/2492494.2492500. 145

[197] Noser, H., Renault, O., Thalmann, D., and Magnenat-Thalmann, N. (1995). Navigation for digital actors based on synthetic vision, memory and learning. *Computers and Graphics*, 19:7–19. DOI: 10.1016/0097-8493(94)00117-H. 145

[198] Nuxoll, A. M. (2007). *Enhancing Intelligent Agents with Episodic Memory*. PhD thesis, Ann Arbor, MI, USA. AAI3287596. DOI: 10.1016/j.cogsys.2011.10.002. 146

[199] Oliva, R. and Pelechano, N. (2011). Automatic generation of suboptimal navmeshes. In *Motion in Games*, Lecture Notes in Computer Science, pages 328–339, Berlin, Heidelberg. Springer-Verlag. DOI: 10.1007/978-3-642-25090-3_28. 57, 67

[200] Ondřej, J., Pettré, J., Olivier, A.-H., and Donikian, S. (2010). A synthetic-vision based steering approach for crowd simulation. *ACM Trans. on Graphics*, 29(4):123:1–123:9. DOI: 10.1145/1778765.1778860. 8, 93

[201] Orkin, J. (2002). Applying goal oriented action planning in games. In *AI Game Programming Wisdom 2*, pages 217–229. Charles River Media. 202

[202] O'Sullivan, C. and Ennis, C. (2011). Metropolis: multisensory simulation of a populated city. In *Intl. Conf. on Games and Virtual Worlds for Serious Applications*, pages 1–7. IEEE Computer Society. DOI: 10.1109/VS-GAMES.2011.9. 105

[Palmer] Palmer, M. Proposition bank. `http://verbs.colorado.edu`. Online; accessed November 2014. 132

[Palmer and Kipper] Palmer, M. and Kipper, K. Verbnet: A class-based verb lexicon. `http://verbs.colorado.edu`. Online; accessed November 2014. 132

[205] Paris, S., Pettré, J., and Donikian, S. (2007). Pedestrian reactive navigation for crowd simulation: a predictive approach. *Computer Graphics Forum*, 26(3):665–674. DOI: 10.1111/j.1467-8659.2007.01090.x. 8

[206] Paul, R., Charles, D., McNeill, M., and McSherry, D. (2011). Adaptive storytelling and story repair in a dynamic environment. In *Interactive Digital Storytelling*, ICIDS'11, pages 128–139, Berlin, Heidelberg. Springer-Verlag. DOI: 10.1007/978-3-642-25289-1_14. 206

[207] Pelechano, N., Allbeck, J. M., and Badler, N. I. (2007). Controlling individual agents in high-density crowd simulation. In *ACM SIGGRAPH/Eurographics symposium on Computer animation*, SCA '07, pages 99–108, Aire-la-Ville, Switzerland, Switzerland. Eurographics Assoc. DOI: 10.1145/1272690.1272705. 7, 23, 56, 168

[208] Pelechano, N., Allbeck, J. M., and Badler, N. I. (2008). *Virtual Crowds: Methods, Simulation, and Control.* Synthesis Lectures on Computer Graphics and Animation. Morgan & Claypool Publishers. DOI: 10.2200/S00123ED1V01Y200808CGR008. 55, 76, 159, 173, 177

[209] Pelechano, N., O'Brien, K., Silverman, B., and Badler, N. I. (2005). Crowd simulation incorporating agent psychological models, roles and communication. In *First Intl. Workshop on Crowd Simulation*, pages 21–30. 145

[210] Pelechano, N., Spanglang, B., and Beacco, A. (2011). Avatar Locomotion in Crowd Simulation. *Intl. Journal of Virtual Reality*, 10:13–19. 23

[211] Pellens, B., De Troyer, O., Bille, W., Kleinermann, F., and Romero, R. (2005). An ontology-driven approach for modeling behavior in virtual environments. In Meersman, R., Tari, Z., and Herrero, P., editors, *On the Move to Meaningful Internet Systems 2005: OTM 2005 Workshops*, volume 3762 of *Lecture Notes in Computer Science*, pages 1215–1224. Springer Berlin Heidelberg. DOI: 10.1007/11575863. 95

[212] Perlin, K. and Goldberg, A. (1996). Improv: A system for scripting interactive actors in virtual worlds. In *ACM SIGGRAPH Conf. on Computer Graphics and Interactive Techniques*, SIGGRAPH '96, pages 205–216, New York, NY, USA. ACM. DOI: 10.1145/237170.237258. 177

[213] Peters, C. (2005). Direction of attention perception for conversation initiation in virtual environments. In Panayiotopoulos, T., Gratch, J., Aylett, R., Ballin, D., Olivier, P., and Rist, T., editors, *Intelligent Virtual Agents*, volume 3661 of *Lecture Notes in Computer Science*, pages 215–228. Springer Berlin Heidelberg. DOI: 10.1007/11550617. 94

[214] Peters, C., Castellano, G., Rehm, M., André, E., Raouzaiou, A., Rapantzikos, K., Karpouzis, K., Volpe, G., Camurri, A., and Vasalou, A. (2011). Fundamentals of agent perception and attention modelling. In Cowie, R., Pelachaud, C., and Petta, P., editors, *Emotion-Oriented Systems*, Cognitive Technologies, pages 293–319. Springer Berlin Heidelberg. 93

[215] Peters, C. and O'Sullivan, C. (2002). Synthetic vision and memory for autonomous virtual humans. *Computer Graphics Forum*, 21(4):743–752. DOI: 10.1111/1467-8659.00632. 145

[216] Pettré, J., Kallmann, M., and Lin, M. C. (2008). Motion planning and autonomy for virtual humans. In *ACM SIGGRAPH Course*, pages 1–31. 58

[217] Pettré, J. and Laumond, J.-P. (2006). A motion capture-based control-space approach for walking mannequins: Research articles. *Computer Animation and Virtual Worlds*, 17(2):109–126. DOI: 10.1002/cav.76. 9

[218] Pettré, J., Laumond, J.-P., and Siméon, T. (2003). A 2-stages locomotion planner for digital actors. In *ACM SIGGRAPH/Eurographics Symp. on Computer Animation*, SCA '03, pages 258–264, Aire-la-Ville, Switzerland, Switzerland. Eurographics Assoc. 9

[219] Pettré, J., Ondřej, J., Olivier, A.-H., Cretual, A., and Donikian, S. (2009). Experiment-based modeling, simulation and validation of interactions between virtual walkers. In *ACM SIGGRAPH/Eurographics Symp. on Computer Animation*, SCA '09, pages 189–198, New York, NY, USA. ACM. DOI: 10.1145/1599470.1599495. 8

[220] Pieper, D. L. (1968). The kinematics of manipulators under computer control. Technical Report STAN-CS-68-116, Department of computer science, Stanford University. 190

[221] Pittarello, F. and De Faveri, A. (2006). Semantic description of 3d environments: A proposal based on web standards. In *Intl. Conf. on 3D Web Technology*, Web3D '06, pages 85–95, New York, NY, USA. ACM. DOI: 10.1145/1122591.1122603. 95

[Princeton University] Princeton University. Wordnet. `http://wordnet.princeton.edu`. Online; accessed November 2014. 132

[223] Quinlan, J. R. (1986). Induction of decision trees. *Machine Learning*, 1(1):81–106. DOI: 10.1023/A:1022643204877. 43

[224] Raghuvanshi, N., Snyder, J., Mehra, R., Lin, M., and Govindaraju, N. (2010). Precomputed wave simulation for real-time sound propagation of dynamic sources in complex scenes. *ACM Trans. on Graphics*, 29(4):68:1–68:11. DOI: 10.1145/1778765.1778805. 113

[225] Reynolds, C. W. (1987). Flocks, herds and schools: A distributed behavioral model. *SIGGRAPH Computer Graphics*, 21(4):25–34. DOI: 10.1145/37402.37406. 7, 23

[226] Reynolds, C. W. (1999). Steering behaviors for autonomous characters. In *Game Developers Conf.*, GDC '99, pages 763–782, San Francisco, California. Miller Freeman Game Group. 7, 56

[227] Rickel, J. and Johnson, W. L. (1999). Animated agents for procedural training in virtual reality: Perception, cognition, and motor control. *Applied Artificial Intelligence*, 13(4-5):343–382. DOI: 10.1080/088395199117315. 145

[228] Riedl, M., Saretto, C. J., and Young, R. M. (2003). Managing interaction between users and agents in a multi-agent storytelling environment. In *Intl. Joint Conf. on Autonomous Agents and Multiagent Systems*, AAMAS '03, pages 741–748, New York, NY, USA. ACM. DOI: 10.1145/860575.860694. 178

[229] Riedl, M. O. and Bulitko, V. (2013). Interactive narrative: An intelligent systems approach. *AI Magazine*, 34(1):67–77. 178

[230] Rodriguez, S. and Amato, N. M. (2011). Roadmap-based level clearing of buildings. In *Motion in Games*, Lecture Notes in Computer Science, pages 340–352, Berlin, Heidelberg. Springer-Verlag. DOI: 10.1007/978-3-642-25090-3_29. 55

[231] Russell, S. J., Norvig, P., Canny, J. F., Malik, J. M., and Edwards, D. D. (1995). *Artificial Intelligence: A Modern Approach*, volume 2. Prentice Hall, Englewood Cliffs, NJ. 132

[232] Rymill, S. J. and Dodgson, N. A. (2005). Psychologically-based vision and attention for the simulation of human behaviour. In *Intl. Conf. on Computer Graphics and Interactive Techniques in Australasia and South East Asia*, GRAPHITE '05, pages 229–236, New York, NY, USA. ACM. DOI: 10.1145/1101389.1101435. 94

[233] Sacerdoti, E. D. (1975). The nonlinear nature of plans. In *Intl. Joint Conf. on Artificial Intelligence*, pages 206–214. 178

[234] Safonova, A. and Hodgins, J. K. (2007). Construction and optimal search of interpolated motion graphs. *ACM Trans. on Graphics*, 26(3). DOI: 10.1145/1276377.1276510. 58

[235] Savioja, L., Huopaniemi, J., Lokki, T., and Väänänen, R. (1999). Creating interactive virtual acoustic environments. *Journal of the Audio Engineering Society*, 47(9):675–705. 94, 141

[236] Schacter, D. L. (2001). *The Seven Sins of Memory*. Houghton Mifflin. 151

[237] Scherer, S., Marsella, S., Stratou, G., Xu, Y., Morbini, F., Egan, A., Rizzo, A., and Morency, L.-P. (2012). Perception markup language: Towards a standardized representation of perceived nonverbal behaviors. In Nakano, Y., Neff, M., Paiva, A., and Walker, M., editors, *Intelligent Virtual Agents*, volume 7502 of *Lecture Notes in Computer Science*, pages 455–463. Springer Berlin Heidelberg. DOI: 10.1007/978-3-642-33197-8. 94

[238] Schneider, J., Garatly, D., Srinivasan, M., Guy, S. J., Curtis, S., Cutchin, S., Manocha, D., Lin, M. C., and Rockwood, A. (2011). Towards a Digital Makkah—Using Immersive 3D Environments to Train and Prepare Pilgrims. In *Digital Media and its Applications in Cultural Heritage*, pages 1–16. 7

[239] Schuerman, M., Singh, S., Kapadia, M., and Faloutsos, P. (2010). Situation agents: agent-based externalized steering logic. *Computer Animation and Virtual Worlds*, 21:267–276. DOI: 10.1002/cav.367. 51

[240] Shapiro, A. (2011). Building a character animation system. In Allbeck, J. and Faloutsos, P., editors, *Motion in Games*, volume 7060 of *Lecture Notes in Computer Science*, pages 98–109. Springer-Verlag, Berlin, Heidelberg. 190

[241] Shapiro, A., Kallmann, M., and Faloutsos, P. (2007). Interactive motion correction and object manipulation. In *ACM SIGGRAPH Symp. on Interactive 3D Graphics and Games*, SI3D'07, pages 137–144. DOI: 10.1145/1230100.1230124. 58

[242] Shoulson, A. and Badler, N. I. (2011). Event-centric control for background agents. In *Intl. Conf. on Interactive Digital Storytelling*, ICIDS'11, pages 193–198. DOI: 10.1007/978-3-642-25289-1_21. 177

[243] Shoulson, A., Garcia, F., Jones, M., Mead, R., and Badler, N. I. (2011). Parameterizing behavior trees. In *Motion in Games*, Lecture Notes in Computer Science, pages 144–155, Berlin, Heidelberg. Springer-Verlag. DOI: 10.1007/978-3-642-25090-3_13. 184, 194, 200

[244] Shoulson, A., Kapadia, M., and Badler, N. I. (2013a). Paste: A platform for adaptive storytelling with events. *Intelligent Narrative Technologies*, 6. 216

[245] Shoulson, A., Marshak, N., Kapadia, M., and Badler, N. I. (2013b). Adapt: The agent development and prototyping testbed. In *ACM SIGGRAPH Symp. on Interactive 3D Graphics and Games*, I3D '13, pages 9–18, New York, NY, USA. ACM. DOI: 10.1145/2448196.2448198. 29, 217

[246] Shoulson, A., Marshak, N., Kapadia, M., and Badler, N. I. (2013c). ADAPT: the agent development and prototyping testbed. In *ACM SIGGRAPH Symp. on Interactive 3D Graphics and Games*, I3D '13, pages 9–18. DOI: 10.1145/2448196.2448198. 112, 179

[247] Shum, H. P. H., Komura, T., Shiraishi, M., and Yamazaki, S. (2008). Interaction patches for multi-character animation. In *ACM SIGGRAPH Asia*, SA'08, pages 114:1–114:8. DOI: 10.1145/1409060.1409067. 177

[248] Si, M., Marsella, S. C., and Pynadath, D. V. (2005). THESPIAN: An architecture for interactive pedagogical drama. In *Artificial Intelligence in Education*, pages 595–602. 178

[249] Silverman, B. G., Johns, M., Cornwell, J., and O'Brien, K. (2006). Human behavior models for agents in simulators and games: Part I: Enabling science with PMFserv. *Presence: Teleoperators and Virtual Environments*, 15(2):139–162. DOI: 10.1162/pres.2006.15.2.163. 169

[250] Singh, S., Kapadia, M., Faloutsos, P., and Reinman, G. (2009a). An open framework for developing, evaluating, and sharing steering algorithms. In *Motion in Games*, Lecture Notes in Computer Science, pages 158–169, Berlin, Heidelberg. Springer-Verlag. DOI: 10.1007/978-3-642-10347-6_15. 8, 217

[251] Singh, S., Kapadia, M., Faloutsos, P., and Reinman, G. (2009b). SteerBench: a benchmark suite for evaluating steering behaviors. *Computer Animation and Virtual Worlds*, 9999(9999):n/a+. DOI: 10.1002/cav.277. 46, 217

[252] Singh, S., Kapadia, M., Hewlett, B., Reinman, G., and Faloutsos, P. (2011a). A modular framework for adaptive agent-based steering. In *ACM SIGGRAPH Symp. on Interactive 3D Graphics and Games*, volume 1 of *I3D'11*, pages 141–150. DOI: 10.1145/1944745.1944769. 8

[253] Singh, S., Kapadia, M., Reinman, G., and Faloutsos, P. (2011b). Footstep navigation for dynamic crowds. *Computer Animation and Virtual Worlds*, 22(2-3):151–158. DOI: 10.1002/cav.403. 9, 15, 23, 33, 41, 58

[254] Singh, S., Kapadia, M., Reinman, G., and Faloutsos, P. (2011c). Footstep navigation for dynamic crowds. In *ACM SIGGRAPH Symp. on Interactive 3D Graphics and Games*, I3D '11, pages 203–203, New York, NY, USA. ACM. DOI: 10.1002/cav.403. 15

[255] Singh, S., Naik, M., Kapadia, M., Faloutsos, P., and Reinman, G. (2008). Watch out! a framework for evaluating steering behaviors. In *Motion in Games*, Lecture Notes in Computer Science, pages 200–209, Berlin, Heidelberg. Springer-Verlag. DOI: 10.1007/978-3-540-89220-5_20. 217

[256] Slonneger, D., Croop, M., Cytryn, J., Kider, Jr., J. T., Rabbitz, R., Halpern, E., and Badler, N. I. (2011). Human model reaching, grasping, looking and sitting using smart objects. In *Intl. Symp. on Digital Human Modeling*. Intl. Ergonomic Assoc. 190

[257] Small, H. (1973). Co-citation in the scientific literature: A new measure of the relationship between two documents. *Journal of American Society for Information Science*, 24(4):265–269. DOI: 10.1002/asi.4630240406. 125

[258] Snook, G. (2000). Simplified 3d movement and pathfinding using navigation meshes. In *Game Programming Gems*, pages 288–304. Charles River Media. 55, 56

[259] Steel, T., Kuiper, D., and Wenkstern, R. Z. (2010). Context-aware virtual agents in open environments. In *Intl. Conf. on Autonomic and Autonomous Systems*, pages 90–96. DOI: 10.1109/ICAS.2010.36. 94, 121

[260] Sternberg, S. (1966). High-speed scanning in human memory. *Science*, 153(3736):652–654. DOI: 10.1126/science.153.3736.652. 152

[261] Stocker, C., Sun, L., Huang, P., Qin, W., Allbeck, J. M., and Badler, N. I. (2010a). Smart events and primed agents. In *Intelligent Virtual Agents*, IVA'10, pages 15–27. Springer-Verlag. DOI: 10.1007/978-3-642-15892-6_2. 177

[262] Stocker, C., Sun, L., Huang, P., Qin, W., Allbeck, J. M., and Badler, N. I. (2010b). Smart events and primed agents. In *Intelligent Virtual Agents*, IVA'10, pages 15–27. Springer-Verlag. DOI: 10.1007/978-3-642-15892-6_2. 194

[263] Sturtevant, N. R. and Geisberger, R. (2010). A comparison of high-level approaches for speeding up pathfinding. In *Artificial Intelligence and Interactive Digital Entertainment*, AIIDE'10, pages 76–82. 58

[264] Sud, A., Gayle, R., Andersen, E., Guy, S., Lin, M., and Manocha, D. (2007a). Real-time navigation of independent agents using adaptive roadmaps. In *ACM symposium on Virtual Reality Software and Technology*, VRST '07, pages 99–106, New York, NY, USA. ACM. DOI: 10.1145/1315184.1315201. 55, 56

[265] Sud, A., Gayle, R., Andersen, E., Guy, S., Lin, M., and Manocha, D. (2007b). Real-time navigation of independent agents using adaptive roadmaps. In *ACM Symp. on Virtual Reality Software and Technology*, pages 99–106. ACM. DOI: 10.1145/1315184.1315201. 58

[266] Sun, L., Shoulson, A., Huang, P., Nelson, N., Qin, W., Nenkova, A., and Badler, N. I. (2012). Animating synthetic dyadic conversations with variations based on context and agent attributes. *Computer Animation and Virtual Worlds*, 23(1):17–32. DOI: 10.1002/cav.1421. 195

[267] Sun, R. (2006). The clarion cognitive architecture: Extending cognitive modeling to social simulation. pages 79–100. 146

[268] Sung, M., Kovar, L., and Gleicher, M. (2005). Fast and accurate goal-directed motion synthesis for crowds. In *ACM SIGGRAPH/Eurographics Symp. on Computer Animation*, SCA'05, pages 291–300. DOI: 10.1145/1073368.1073410. 58, 76

[269] Sunshine-Hill, B. and Badler, N. I. (2010). Perceptually realistic behavior through alibi generation. In Youngblood, G. M. and Bulitko, V., editors, *Artificial Intelligence and Interactive Digital Entertainment*, AIIDE'10. The AAAI Press. 217

[270] Takala, T. and Hahn, J. (1992). Sound rendering. *SIGGRAPH Computer Graphics*, 26(2):211–220. DOI: 10.1145/142920.134063. 94

[271] Thalmann, D. and Musse, S. R. (2012). *Crowd Simulation*. Springer, 2 edition. DOI: 10.1007/978-1-4471-4450-2. 7

[272] Thalmann, D. and Musse, S. R. (2013). *Crowd Simulation*. Springer, 2 edition. DOI: 10.1007/978-1-4471-4450-2. 177, 217

[273] Thawonmas, R., Hirayama, J., and Takeda, F. (2002). RoboCup Agent Learning from Observations with Hierarchical Multiple Decision Trees. *PRIMA*. 43

[274] Thomas, R. and Donikian, S. (2007). A spatial cognitive map and a human-like memory model dedicated to pedestrian navigation in virtual urban environments. In Barkowsky, T., Knauff, M., Ligozat, G., and Montello, D. R., editors, *Spatial Cognition V Reasoning, Action, Interaction*, volume 4387 of *Lecture Notes in Computer Science*, pages 421–438. Springer Berlin Heidelberg. DOI: 10.1007/978-3-540-75666-8. 145

[275] Thue, D., Bulitko, V., Spetch, M., and Wasylishen, E. (2007). Interactive storytelling: A player modelling approach. In *Artificial Intelligence and Interactive Digital Entertainment Conf.*, AIIDE'07. 178

[276] Torkos, N. and van de Panne, M. (1998). Footprint-based quadruped motion synthesis. In *Graphics Interface*, GI'98, pages 151–160. Canadian Human-Computer Communications Society. 14

[277] Torrens, P., Li, X., and Griffin, W. A. (2011). Building Agent-Based Walking Models by Machine-Learning on Diverse Databases of Space-Time Trajectory Samples. *Trans. in GIS*, 15:67–94. DOI: 10.1111/j.1467-9671.2011.01261.x. 8, 40

[278] Tozour, P. (2002). Ai game programming wisdom. In Rabin, S., editor, *Building a Near-Optimal Navigation Mesh*, pages 171–185. Charles River Media. 57

[279] Treuille, A., Cooper, S., and Popović, Z. (2006). Continuum crowds. *ACM Trans. on Graphics*, 25(3):1160. DOI: 10.1145/1141911.1142008. 7, 23

[280] Treuille, A., Lee, Y., and Popović, Z. (2007). Near-Optimal Character Animation with Continuous Control. *ACM Trans. on Graphics*, 26(3):7. DOI: 10.1145/1276377.1276386. 9

[281] Tulving, E. (1983). *Elements of episodic memory*. Clarendon Press. 145

[282] Turetsky, R. J. and Ellis, D. P. W. (2003). Ground-truth transcriptions of real music from force-aligned MIDI syntheses. *ISMIR 2003*, pages 135–141. 103, 109

[283] Tutenel, T., Bidarra, R., Smelik, R. M., and Kraker, K. J. D. (2008). The role of semantics in games and simulations. *Computers in Entertainment*, 6(4):57:1–57:35. DOI: 10.1145/1461999.1462009. 95, 131, 132

[284] UDK (2015). Unreal NavMesh Generator. `http://udn.epicgames.com/Three/NavigationMeshReference.html`. 57

[285] Unity (2014). Unity game engine. `http://unity3d.com/`. 33

[286] Valve (2005). Valve NavMesh Generator. `http://developer.valvesoftware.com/wiki/Navigation_Meshes`. 56

[287] van Basten, B. J. H., Peeters, P. W. A. M., and Egges, A. (2010). The Step Space: Example-Based Footprint-Driven Motion Synthesis. *Computer Animation and Virtual Worlds*, 21(May):433–441. DOI: 10.1002/cav.342. 9

[288] van Basten, B. J. H., Stüvel, S. A., and Egges, A. (2011). A Hybrid Interpolation Scheme for Footprint-Driven Walking Synthesis. In *Graphics Interface*, GI'11, pages 9–16. 9, 24

[289] van de Panne, M. (1997). From Footprints to Animation. *Computer Graphics Forum*, 16(4):211–223. DOI: 10.1111/1467-8659.00181. 9, 14

[290] van den Berg, J., Ferguson, D., and Kuffner, J. (2006). Anytime path planning and replanning in dynamic environments. In *IEEE Intl. Conf. on Robotics and Automation*, ICRA'06, pages 2366 –2371. DOI: 10.1109/ROBOT.2006.1642056. 58

[291] van den Berg, J., Patil, S., Sewall, J., Manocha, D., and Lin, M. (2008a). Interactive navigation of multiple agents in crowded environments. In *ACM SIGGRAPH Symp. on Interactive 3D Graphics and Games*, I3D'08, pages 139–147. DOI: 10.1145/1342250.1342272. 58

[292] van den Berg, J. P., Lin, M. C., and Manocha, D. (2008b). Reciprocal velocity obstacles for real-time multi-agent navigation. In *IEEE Intl. Conf. on Robotics and Automation*, ICRA'08, pages 1928–1935. DOI: 10.1109/ROBOT.2008.4543489. 8

[293] van Oijen, J. and Dignum, F. (2011). Scalable perception for BDI-agents embodied in virtual environments. In *IEEE/WIC/ACM Intl. Conf. on Web Intelligence and Intelligent Agent Technology*, volume 2 of *WI-IAT*, pages 46–53. DOI: 10.1109/WI-IAT.2011.176. 93, 94, 128

[294] Van Rossum, G. (2007). Python programming language. USENIX Annual Technical Conf. 133

[295] van Toll, W., Cook IV, A. F., and Geraerts, R. (2012a). A navigation mesh for dynamic environments. *Journal of Visualization and Computer Animation*, 23(6):535–546. DOI: 10.1002/cav.1468. 57

[296] van Toll, W. G., Cook IV, A. F., and Geraerts, R. (2012b). Real-time density-based crowd simulation. *Computer Animation and Virtual Worlds*, 23(1):59–69. DOI: 10.1002/cav.1424. 78

[297] Vilhjálmsson, H., Cantelmo, N., Cassell, J., Chafai, N., Kipp, M., Kopp, S., Mancini, M., Marsella, S., Marshall, A., Pelachaud, C., Ruttkay, Z., Thórisson, K. R., Welbergen, H., and Werf, R. (2007). The behavior markup language: Recent developments and challenges. In *Intelligent Virtual Agents*, IVA '07, pages 99–111. DOI: 10.1007/978-3-540-74997-4_10. 190

[298] Wagemans, J., Elder, J. H., Kubovy, M., Palmer, S. E., Peterson, M. A., Singh, M., and von der Heydt, R. (2012). A century of Gestalt psychology in visual perception: I. Perceptual grouping and figure-ground organization. *Psychological Bulletin*, 138(6):1172–1217. DOI: 10.1037/a0029334. 124

[299] Wang, Y., Kapadia, M., Huang, P., Kavan, L., and Badler, N. I. (2014). Sound localization and multi-modal steering for autonomous virtual agents. In *ACM SIGGRAPH Symp. on Interactive 3D Graphics and Games*, I3D '14, New York, NY, USA. ACM. DOI: 10.1145/2556700.2556718. 141

[300] Williams, H. L., Conway, M. A., and Cohen, G. (2008). Autobiographical memory. In Cohen, G. and Conway, M. A., editors, *Memory in the real world*, Lecture Notes in Computer Science, pages 21–90. Psychology Press, London, 3 edition. 145

[301] Won, J., Lee, K., O'Sullivan, C., Hodgins, J. K., and Lee, J. (2014). Generating and ranking diverse multi-character interactions. *ACM Trans. on Graphics*, 33(6):219:1–219:12. DOI: 10.1145/2661229.2661271. 177

[302] Xu, C., Maddage, N. C., and Shao, X. (2005). Automatic music classification and summarization. *IEEE Trans. on Speech and Audio Processing*, 13(3):441–450. DOI: 10.1109/TSA.2004.840939. 99

[303] Yeo, S. H., Lesmana, M., Neog, D. R., and Pai, D. K. (2012). Eyecatch: Simulating visuomotor coordination for object interception. *ACM Trans. on Graphics*, 31(4):42:1–42:10. DOI: 10.1145/2185520.2185538. 94

[304] Young, R. M. and Laird, J. E., editors (2005). *Artificial Intelligence and Interactive Digital Entertainment Conf.* AAAI Press. 177

[305] Young, T. (2001). Expanded geometry for points-of-visibility pathfinding. In *Game Programming Gems 2*, pages 317–323. Charles River Media. 55

[306] Yu, Q. and Terzopoulos, D. (2007). A decision network framework for the behavioral animation of virtual humans. In *ACM SIGGRAPH/Eurographics Symp. on Computer Animation*, SCA '07, pages 119–128, Aire-la-Ville, Switzerland, Switzerland. Eurographics Assoc. DOI: 10.1145/1272690.1272707. 177

Authors' Biographies

MUBBASIR KAPADIA

Mubbasir Kapadia is an Assistant Professor in the Computer Science Department at Rutgers University. Previously, he was an Associate Research Scientist at Disney Research Zurich. He was a postdoctoral researcher and Assistant Director at the Center for Human Modeling and Simulation at University of Pennsylvania, under the directorship of Prof. Norman I. Badler. He was the project lead on the United States Army Research Laboratory (ARL) funded project Robotics Collaborative Technology Alliance (RCTA). He received his Ph.D. in Computer Science at University of California, Los Angeles, under the advisement of Professor Petros Faloutsos.

NURIA PELECHANO

Nuria Pelechano is an Associate Professor at the Universitat Politècnica de Catalunya. She obtained her Engineering degree from the Universitat de Valencia, her Masters degree from the University College London, and her Ph.D. from the University of Pennsylvania as a Fulbright Scholar in 2006. Nuria has over 30 publications in journals and international conferences on Computer Graphics and Animation. She has participated in projects funded by the EU, the Spanish Government, and U.S. institutions. Her research interests include simulation, animation and rendering of crowds, generation of navigation meshes, real-time 3D graphics, and human-avatar interaction in virtual environments.

JAN M. ALLBECK

Jan M. Allbeck is an Associate Professor in the Department of Computer Science at George Mason University, where she is also the faculty advisor for their undergraduate concentration in Computer Game Design and director of the Games and Intelligent Animation laboratory. She received her Ph.D. in Computer and Information Science from the University of Pennsylvania in 2009. She has more than 50 publications in international journals and conference proceedings and has served as a reviewer for 40 journals, conferences, and workshops. She has had the great opportunity to explore many aspects of computer graphics, but is most drawn to research at the crossroads of animation, artificial intelligence, and psychology in the simulation of virtual humans.

NORMAN I. BADLER

Norman I. Badler is Rachleff Professor of Computer and Information Science at the University of Pennsylvania. He received his B.A. in Creative Studies Mathematics from the University of California Santa Barbara in 1970, his MSc in Mathematics from the University of Toronto in 1971, and his Ph.D. in Computer Science from the University of Toronto in 1975. He served as the Senior Co-Editor for the journal *Graphical Models* for 20 years and presently serves on the editorial boards of several other journals, including *Presence*. His research involves developing software to acquire, simulate, animate, and control 3D computer graphics human body, face, gesture, locomotion, and manual task motions, both individually and for heterogeneous groups. He has supervised or co-supervised 62 Ph.D. students, many of whom have become academics or researchers in the movie visual effects and game industries. He is the founding Director of the SIG Center for Computer Graphics, the Center for Human Modeling and Simulation, and the ViDi Center for Digital Visualization at Penn. He has served Penn as Chair of the Computer & Information Science Department (1990–94) and as the Associate Dean of the School of Engineering and Applied Science (2001–05).

Printed in the United States
by Baker & Taylor Publisher Services